THE BLACK HOLE

at the Center of Our Galaxy

THE BLACK HOLE
at the Center of Our Galaxy

Fulvio Melia

✐ PRINCETON UNIVERSITY PRESS

PRINCETON AND OXFORD

Copyright © 2003 by Princeton University Press
Published by Princeton University Press, 41 William
Street, Princeton, New Jersey 08540
In the United Kingdom: Princeton University Press,
3 Market Place, Woodstock, Oxfordshire OX20 1SY

Library of Congress Cataloging-in-Publication Data
Melia, Fulvio.
The black hole at the center of our galaxy / Fulvio Melia.
p. cm.
Includes bibliographical references and index.
ISBN 0-691-09505-1 (acid-free paper)
1. Black holes (Astronomy) 2. Galactic center. I. Title.
QB843.B55 M45 2003
523.8′875—dc21   2002034625

British Library Cataloging-in-Publication Data is available

This book has been composed in Times Roman

The publisher would like to acknowledge the author
of this volume for providing the camera-ready copy
from which this book was printed.

Printed on acid-free paper. ∞
www.pupress.princeton.edu
Printed in the United States of America
10 9 8 7 6 5 4 3 2 1

CONTENTS

PREFACE

On a clear summer night you can see the Milky Way arching southward across the sky. Just before it touches the horizon, at the constellation known as Sagittarius, the archer, a clustering of stars marks the center of this magnificent structure. Astronomers have had difficulty looking into the heart of the Milky Way because our solar system lies in the plane of the galaxy, where thick dust lanes and clouds block our view. But this has changed dramatically in recent years due to our ability to peer through these dusty veils with new telescopes that detect radiation emitted at wavelengths significantly longer and shorter than those of visible light, which is severely attenuated because of scattering by the intervening gas. What we are discovering with the new imagery is that, contrary to expectations, the galactic center is a bizarre place with jets and wisps of very energetic matter, with imploding stars and exploding shells of hot gas, and that at the heart of it all lies the most enigmatic object yet known—a dark entity with the mass of almost 3 million Suns, devouring everything around it.

That the center of the Milky Way is positioned within the constellation of Sagittarius had been established from dynamical evidence many decades ago. When we measure the velocity of stars and the swirling gas clouds surrounding the nucleus, we uncover a velocity profile with a well-defined center in this region, much as the Sun acts as a centroid for the motion of the planets within our solar system. Recent observations, however, have taken us with unprecedented leaps into the inner one-twentieth of a light-year, and we now have enough data to be able to piece together what the center of our galaxy actually looks like. Given that we are viewing this region from a distance of 28,000 light-years, having the ability to distinguish features separated by only one-twentieth of a light-year is comparable to seeing two cars parked adjacent to each other in San Francisco from a vantage point in New York.

One of the most exciting recent developments in this arena has been the realization that the black hole (known as Sagittarius A*) at the galactic center is close enough to us that we should be able to actually image its "surface," referred to as its event horizon by astrophysicists, against the backdrop of nearby, radiating gas—a concept that was completely unimaginable just a matter of years ago. Perhaps the most enigmatic objects in the cosmos, black holes enclose regions of space within which gravity is so strong that not even light can escape—hence their name. Light paths originating outside Sagittarius A* are bent by its strong gravitational field, so that radiation produced by the infalling gas behind it is simply absorbed on its way to us. The size of the shadow produced by this effect is predicted to be well within the resolving power of radio telescopes being developed during this decade. Thus, the galactic center may soon provide us with the means of directly testing the most captivating prediction of general relativity—the existence of black holes.

Of no less interest is the question of what happens when a star ventures too close to such an object. Not unreasonably, we might expect some of the most spectacular fireworks in the sky. Indeed, the gossamer structure seen surrounding the nucleus of our galaxy may be the remnant of such an explosion that took place within the past 100,000 years. This feature looks very much like the expanding shell left over after a supernova explosion—the catastrophic collapse of a dying star after it has spent all of its nuclear fuel—and we can gauge how much energy was liberated, and how long it took to do so, based on how difficult it was for the expanding gas to carve out a cavity in the surrounding molecular cloud. But this explosion was no ordinary supernova. The energy liberated during the explosion was at least 50 to 100 times greater, and it therefore rules out any such simple interpretation. The explosion that produced this expanding shell was instead probably due to the powerful disruptive effects of the black hole's gravitational field, which squeezed the doomed star into a long thin spike during its inward trajectory. When receding from its location of closest approach to the black hole, the stellar debris expanded explosively, very much like a supernova shell, except with a much greater power. The consequences of these events can be quite stunning, as we have seen recently with space-based instruments, such as the Energetic Gamma-Ray Experiment Telescope (EGRET) and the Oriented Scintillation Spectrometer Experiment (OSSE) on the Compton Gamma-Ray Observatory.

The nature of our galactic center is under intense study today. The relationship between the various components in and around the massive black hole Sagittarius A* is still not completely understood. For example, there are filamentary loops stretching out some 400 light-years from the galactic plane that remain something of a mystery. It has been suggested that these threads may be related to "cosmic strings," which are thought to be remnants of the early universe, and which are presumed to help nucleate galaxy formation.

In any case, at this point we are reasonably sure that the galactic center contains a supermassive black hole, because we know of nothing else that could account for the highly peaked, dark matter concentration measured there. Our knowledge of the galactic center will grow rapidly in the coming decade, as more telescopes are devised to observe the full electromagnetic spectrum and as more of the cosmos is studied with sensors that are unobstructed by Earth's atmosphere. And one of the most important discoveries since the development of relativity in the early part of the twentieth century—the imaging of an event horizon—is likely to occur during this period.

These are among the elements that have motivated the writing of this book. The exciting discoveries produced by the vibrant galactic center research community are at times breathtaking and ought to be made accessible to a general audience. For my part, I have had the good fortune over the years to assist in the exploration of the central region of our galaxy, and in these endeavors I am very grateful to my students and collaborators for the pleasure of our joint efforts. They include Peter Tamblyn, Jack Hollywood, Laird Close, Sera Markoff, Alexei Khokhlov, Marco Fatuzzo, Robert Coker, Mike Fromerth, Siming Liu, Gabe Rockefeller, Susan Stolovy, Heino Falcke, Victor Kowalenko, Ray Volkas, Roland Crocker, Pasquale Blasi, Don MacCarthy, George Rieke, Benjamin Bromley, Randy Jokipii, and Martin Pessah.

I am particularly grateful to my close friend and long-time collaborator, Farhad Yusef-Zadeh, whose early radio images of the galactic center inspired my interest in this field, and whose ongoing drive and ground-breaking observations continue to be a fountain of enthusiasm and new ideas. I am indebted also to Timothy Ferris and Sir Martin Rees, who graciously took the time to read the manuscript and offer their valued expert advice on ways to improve the presentation. A wise and experienced editor, Joe Wisnovsky has been an invaluable resource throughout the course of this project, and I have benefited greatly from his guidance and encouragement. And for generously supporting my research in this area for over a decade and a half, I gratefully acknowledge the National Science Foundation, the National Aeronautics and Space Administration, and the Alfred P. Sloan Foundation.

Finally, I owe a debt of gratitude to the pillars of my life— Patricia, Marcus, Eliana, and Adrian—and to my parents, whose guidance has been priceless.

Fulvio Melia
Tucson, Arizona

THE BLACK HOLE
at the Center of Our Galaxy

1

THE GALACTIC CENTER

Take a trip some day to the coastal town of Port Douglas on the northeastern shore of Australia. There, lying on the warm, sandy beach, you will witness the evening splendor of the Milky Way arching from horizon to horizon across the southern cosmic vault (figure 1.1). And as your eyes peer into the starry void, you will marvel at the serene magnificence of this beautiful and overpowering structure that we proudly call *our* galaxy. But succumbing as you do to its hypnotic spell, you will be unaware that its center conceals another universe, shielded from us by a one-way membrane—an event horizon—that eternally forbids the world within it to make contact with the physical reality you will be sensing around you on this night of enchantment.

As a species accustomed to seeking truth and finding beauty in the heart of all things, we find ourselves beckoned by the primacy of the central realm. In Jules Verne's science fiction classic *A Journey to the Center of the Earth*, Professor Hardwigg and his fellow explorers encounter an assortment of strange, breathtaking wonders as they approach Earth's core. So it should be, we believe, for other than at the nucleus of this planet, where else ought there be a place more special, more endowed with the qualities that make something unique? Throughout history, it seems, humans have looked to the center of their environment for privilege, wisdom, and comfort. In early Chinese culture, art and invention were to be found only in the "central kingdom," a

Figure 1.1 Obscured by ragged, dark dust clouds, the galactic center is virtually unobservable even with the largest optical telescopes. This image, made in Cerro Tololo, Chile, shows the region near the border between the constellations of Scorpius and Sagittarius; the former is partially visible at the bottom right, and the latter to the left, in the photograph. North is at the top and West is to the right in this view, which spans an area of about 11,000 by 17,000 light-years. The dynamical center of the galaxy coincides with the cross-mark on the diagonal line that cuts through the galactic plane. (Photograph courtesy of William Keel at the University of Alabama, Tuscaloosa)

sentiment echoed with power by the Romans in defense of their imperial capital city of Rome, the center of anti-barbarism. By extension, the heart of something as majestic as the Milky Way must be special indeed, and we are drawn to it, teased by what it reveals, tormented with the desire to see more.

Our intellectual pilgrimage to the center of the galaxy traces its roots to embryonic attempts in the mid-twentieth century to view the heavens with light much redder and much bluer than the human eye can sense. Yet only in recent years have we been granted the privilege of focusing our attention on the nucleus. And what a journey this is turning out to be. We will begin our trek by embarking on a virtual exploration, navigating through a series of progressively higher-resolution images, reaching a detailed view whose unprecedented clarity challenges the imagination. We will learn about the unique players on this stage, and the nature of the most enigmatic object of them all—the supermassive black hole—caged within a pit of strong gravity in the galaxy's most mysterious domain.

1.1   THE HIDDEN REALM

The Milky Way is a giant galaxy, encompassing a mass of almost 1 trillion Suns within a diameter of about 100,000 light-years. (A light-year is the distance light travels in a year. By comparison, it takes light 2 seconds to make a round trip to the moon and back, and eight minutes to reach us from the Sun.) The Milky Way belongs to the so-called Local Group of galaxies, a small clustering of over 30 members, and is the second largest, and one of the most massive representatives of this ensemble. The nearest large galaxy, Andromeda (figure 1.2), is 2.4 million light-years away, though several fainter galaxies are much closer. Among them are the conspicuous Large and Small Magellanic Clouds at a distance of 179,000 and 210,000 light-years, respectively.

Our understanding of how the Milky Way came into existence, and of what forces govern its current evolution, began to take shape in the latter half of the twentieth century, when

Figure 1.2 The Andromeda galaxy is a large spiral, comparable in size and appearance to the Milky Way, were the latter to be viewed from outside. Drifting majestically 2.4 million light-years from Earth, we see it today as it appeared that many years ago. Andromeda is the most distant object visible to the unaided eye. (Photograph courtesy of T. A. Rector, B. A. Wolpa, and NOAO/AURA/NSF)

astronomers developed the ability to determine the age of stars and map their concentration with some precision. Our galaxy, we learned, formed from the rapid collapse of a spherical cloud of matter in the primordial expansion of the universe. Most of the elements heavier than helium are the ashes produced by nuclear burning inside stars—much later in the galaxy's evolution—so the young nursery that spawned the nascent Milky Way was replete with simple forms of matter, particularly hydrogen. During this collapse, pockets of high gas condensation formed, and these eventually turned into hot stellar balls of plasma. Today, we see these stars orbiting the center of the galaxy predominantly within a flattened pancake, though many near the middle conspire to puff out a central bulge, like the exaggerated hub of an old cartwheel.

By mapping the distribution of its hydrogen clouds, radio astronomers have further concluded that the Milky Way not only contains a flattened disk, but that it is in fact a spiral galaxy whose face-on view projects a highly ordered grand design with graceful arms gilded by glowing star clusters and hot, tenuous gases. Viewed edge-on, its appearance would be noticeably different but just as striking, with a bulging central region and a precariously thin disk tapering off into the darkness of intergalactic space. Our view is actually closer to the latter than the former since our Sun is embedded within the plane. Most striking are the "milky," sometimes wispy, bands of light produced by the multitude of stars in this swath, cut by the dark, obscuring dust clouds silhouetted against the local spiral arms. It is believed that our galaxy is not unlike Andromeda in appearance, so that alien eyes peering back toward the Milky Way from large distances probably see it very much as we see its sister galaxy in figure 1.2.

Today, the orbits of stars within the galaxy are presumed to reflect the motion of gas from which they formed, locked in harmony to the rotational imprints from the past. The Milky Way rotates gently, though steadily, on an axis whose direction was ordained by the swirling action of the primeval gas. Based on the age of the oldest stars we see, it appears that our galaxy began life as an identifiable separate entity some 10 billion years ago, roughly twice the age of the Sun. Of course, the combination of processes that governed how the initial collapse occurred are not entirely clear, and are still subject to some debate, but our concept of how this sequence of events is ordered does appear to be correct. Some people question whether the contraction to a flattened distribution happened all at once, and whether the thin halo of gas and stars away from the plane might rather have had a different genesis. Several young stars are bucking the trend established by their milieu, and rotate instead in a direction opposite to that of the rest of the galaxy—an indication that they probably originated from the merger of miniature failed galaxies with the much bigger Milky Way.

The Milky Way, however, is not an archetype for all the galaxies we can see in the universe. Some are indeed thin pancakes, but lack a central bloated hub of stars. Others are not flattened disks at all; rather, they fill the void with spheroidal aggregates of luminous and nonluminous matter, swirling in space like slowly spinning footballs. Astronomers are now finding that there may be an evolutionary link that governs how and when central bulges form, or under more extreme conditions, how they grow to dominate the entire galaxy. Ellipticals themselves may be produced in catastrophic collisions that lead to the eventual merger of two spiral galaxies.

Indeed, the Milky Way and Andromeda, the two dominant spiral galaxies in our Local Group, are falling toward each other at 300,000 miles per hour. Our descendants eons from now will see Andromeda gradually grow in size until, some 3 billion years hence, the two sister galaxies begin to tear at each other's fringes. Eventually, stars from both doomed spirals will plunge past each other, driven by the gravitational force of the two gargantuan galaxies. The Sun itself, together with our planet, will either spin completely out of our galaxy altogether, traveling on a long, desolate path with very few other stars visible in the night sky, or it may plunge toward the center of the newly formed structure where a cacophony of activity will greet it. Since the Sun is expected to burn and sustain life on Earth for another 5 billion years, intelligent life here will see all of this unfold, albeit at a very slow pace. A billion years later, the two beautiful spiral galaxies will have merged into a giant elliptical spheroidal mass of aging stars. Ironically, though, very few stars will actually collide with each other during this encounter, since most of space is filled with wispy gas. Aside from dramatic changes in the appearance of the night sky, our descendants will continue to live on a planet peacefully orbiting around the Sun.

For now, however, the most intriguing region of the cosmos accessible to us appears to be the centralized hub of our galaxy. Ongoing surveys reveal that this may not be a coincidence; we infer from them that only galaxies with a bulge in the middle, or those that are comprised entirely of a spheroidal distribution of stars, harbor a central pit of mysterious dark matter (see chapter 6). Our target in this book—the heart of the Milky Way—lies in the direction of the constellation Sagittarius, close to the border with the neighboring constellation Scorpius (figure 1.1). These two groups of stars, like the other ten constellations in the Zodiac, are probably the oldest patterns recognized by human civilizations. In Greek mythology, Sagittarius is a centaur, a warlike creature with the torso of a man and the body of a horse. This grouping was also known to earlier civilizations in the Middle East; several of those in the Mesopotamian area associated it with their gods of war, variants of the archer-god Nergal. It appears that the identification of this constellation with an archer was universal. Today, we tend to name celestial objects and features after the constellation in which they are found, so the galactic center is said to lie in the Sagittarius A complex, and gaseous structures within it are called, for example, Sagittarius A East and Sagittarius A West. As we will see, the most unusual object in this region, discovered in 1974, stands out on a radio map as a bright dot.[1] It was given the name *Sagittarius A**, the asterisk meant to convey its uniqueness and importance.

The solar system meanders through the outer reaches of the Milky Way, well within the disk and only about 20 light-years above the equatorial symmetry plane, though about 28,000 light-years from the galactic center. Your view of the night sky from Port Douglas will therefore be that of a luminous band splashed across the "galactic equator." Orbiting the center on a nearly circular trajectory, the Sun and its planets move at 250 kilometers per second, and take 220 million years to complete one orbit. By

---

[1] The discovery was made by Balick and Brown (1974), following a prediction by Lynden-Bell and Rees (1971) that such an object might exist at the galactic center.

comparison, our human ancestor *Ardipithecus ramidus* began to split from other specialized forms of hominid species that eventually died out only 5 million years ago. In total, the elements that make up our bodies have orbited the center of the galaxy about 20 times since the Sun was formed roughly 4.6 billion years ago.

So why has it taken so long for anyone to recognize the significance of Sagittarius's battlefield? In large part, the answer is "dust." That ubiquitous and relentless vagrant of the household is often just as pesky for astronomers. Dust in space is chemically simpler and typically smaller than its household cousin by a factor of 10 to 100, but if space dust were allowed to settle on your desk, it would likely appear very much the same as the traditional enemy of the varnished. Dust in space occurs wherever any kind of matter exists in concentration. The solar system is filled with it, and its presence is inferred from a triangular glow that appears near the horizon in the evening after sunset. Known as zodiacal light, this faint illumination is produced when sunlight is reflected by tiny dust particles orbiting the Sun.

Dust also appears in the dark space between the stars, but it is not entirely clear where or how it originates. Almost certainly some (perhaps most) of it is produced in the envelopes of very old, puffed-out stars near the end of their lives. As they expand and shed their tenuous envelope, the material within it cools rapidly and molecules are able to crystallize. Eventually, the crystals grow into dust grains composed of carbon and nitrogen ices.

In retrospect it may seem odd that even into the early 1900s people did not recognize that the easily visible dark lanes in the Milky Way (see figure 1.1) were due to absorption, not structure. They were not bothered by the idea that directly toward the center of the galaxy, whose position was by then fairly well known, there weren't as many stars as just off-center. We now realize that if the galactic nucleus were unobscured, its size and brightness would be comparable to that of the full moon, fading away rapidly into the halo and merging gradually with the stellar disk.

The effect of dust on what we see depends rather sensitively on the color of light we are trying to sense. We will learn much more about the nature of light in chapter 3, but for the moment it suffices to say that in some ways it behaves like waves. Imagine yourself sitting in a gondola on the waters just off Piazza San Marco in Venice. For this analogy, the water is light and the gondola is a dust grain. Water waves that undulate very slowly, so that crests pass by the boat at very long intervals, have little influence; you continue to sit there peacefully, while the waves pass by with virtually no disruption. Waves that have intervals much smaller than the size of the boat—basically ripples—also have little effect; they "bounce" off the gondola with minimal interference. However, waves for which the crests are separated by a distance comparable to the size of the boat will disturb it significantly, and the gondola in turn will disrupt the waves as well.

A very similar effect governs the appearance of the daytime sky. Dust particles in our atmosphere (like those in space) are small, about 0.0001 centimeter across, which happens to be roughly the distance between neighboring crests in visible light. Bluer light has a shorter wavelength (as the crest separation is called) so it behaves more like the ripples on the canal, whereas redder light has a longer wavelength and it passes through the atmosphere virtually unaffected. Sunsets are therefore red, since it is predominantly the red light from the Sun that skims above the horizon through the Earth's thick atmosphere, while the sky is blue, since the shorter-wavelength blue light is scattered about more readily.

It is ironic that the light for which our eyes are best suited to see also happens to have the wavelength for which space dust is the greatest nuisance. Dust dims our view toward the galactic center by a factor of at least 100 million. And so it happens that the heart of the Milky Way, which would otherwise be the brightest patch of nighttime sky, is in fact so heavily obscured by dust that even the most powerful optical telescopes are useless for observing it.

We are fortunate indeed that the development of tools to detect radio waves, following quickly on the heels of the emergence of radar technology in the mid-twentieth century, has opened up bright new vistas in the heavens. Radio waves have a crest separation of a centimeter or more, far greater than the size of dust

grains in space. Like the slowly undulating water waves flowing past the gondola in Venice, they bypass the dust with no
discernible effect, so the experience of looking at the galactic center using a radio telescope is similar in many ways to the feeling of
liberation one gets when the fog lifts and the visibility increases.
We will see that the gloomy appearance of the galactic center on
an optical photograph (see figure 1.1) contrasts sharply with the
fountain of brilliance animating radio images of the same region.

1.2   REMOVING THE DUSTY VEILS

The role of every telescope is to gather as much light as possible
and to focus it onto a detector. Before the advent of electronic
automation, the detector was often a human eye, squinting at
images fed through an array of glass lenses. These days, the detector for such a telescope could very well be a photographic film
(which produced figures 1.1 and 1.2) or a charge-coupled device
(CCD), which works in much the same way as a digital camera.
Radio telescopes operate in a fashion similar to that of optical
devices, though with several notable differences having to do with
the wavelength of the radiation. Visible light has a wavelength of
only about 0.0001 centimeter, so even a modest telescope with a
diameter of 1 meter can create images with eye-pleasing sharpness
given that one can fit a million crests of this light within a mirror
of that size. But light with a wavelength of 1 centimeter would
fill such a device with only 100 crests, producing a much poorer
resolution and an unsatisfying image.

   To achieve the same effect with a radio telescope (see figure 1.3), the mirror—or dish—should therefore be about 10,000
times bigger (roughly 10 kilometers in diameter!). Fortunately,
there are ways around this hurdle, about which we will learn more,
and at any rate, spectacular imagery is feasible even with smaller
radio telescopes (see figure 1.4). This is the reason why the lexi-
con of radio astronomy includes terms such as "radio dish," rather
than lens or mirror, given the immensity of the structure needed to

Figure 1.3 The Very Large Array (VLA) in Socorro, New Mexico, is one of the world's premier astronomical radio observatories. Arranged in a large Y pattern, the 27 antennas span a region up to 36 kilometers (22 miles) across, roughly one and a half times the size of Washington, D.C. Each of the antennas is 25 meters in diameter. However, the signal from all 27 can be combined electronically to yield a resolution of an antenna 36 kilometers across, with the sensitivity of an equivalent dish 130 meters wide. (Image courtesy of NRAO/AUI)

conduct any meaningful observation. You'll know how impressive a radio telescope can be if you've ever stood beneath one—the biggest among them are bigger than a football field. For a radio telescope, the receiver is a combination of a feed antenna at the focus, an amplifier (since the signal is often very weak) and a power detector. It works by measuring the intensity of the radio waves at every point in the patch of sky being studied, and then converting this level of radiation into a color that the human eye can recognize. So figure 1.4, for example, is a map of 90-centimeter

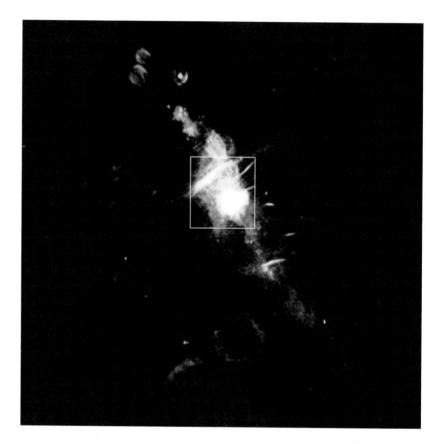

Figure 1.4 Radiation with a wavelength significantly larger than the size of dust particles (which are typically 0.0001 centimeters) in the interstellar medium can pass through the dust clouds without any noticeable attenuation. Thus, if our eyes were sensitive to light with a wavelength of 90 centimeters, the galactic center would reveal itself to us as one of the brightest and most intricate regions of the sky. Fortunately, the Very Large Array (VLA) can do that. This radio image, which spans an area of about 1,000 light-years on each side, i.e., about 1/10 the size of figure 1.1, shows a rich morphology that contrasts with the mostly obscured central region of that earlier optical image. There are no stars visible here because their radiation peaks in the infrared to ultraviolet portion of the spectrum. The radio emission seen here is instead produced by supernova remnants (the circular structures), wispy, snakelike synchrotron filaments, and highly ionized hydrogen gas. The central box shows the bright region magnified in figure 1.5. (Image courtesy of N. E. Kassim and collaborators, Naval Research Laboratory, Washington, D.C.)

radio intensity from the galactic center, though converted into a pre-selected sliding scale of colors that our eyes can interpret. The proper choice of colors preserves our intuition about which regions are bright and which are not.

Jules Verne and Professor Hardwigg would be thrilled with this image, which confirms what our deep-rooted expectations would lead us to believe—a hitherto unexplorable region opens up to reveal a glorious panorama of activity, culminating in a crescendo of light toward (where else?) the middle. This spectacular 1,000 light-year view is the largest and most sensitive radio image ever made of the Milky Way's center at a uniformly high resolution. The concentration of objects and features along the diagonal from the upper left to the lower right reveals the disklike shape of the Milky Way viewed edge-on, and presents a harvest of beauty and complexity to excite the imagination of experts and nonexperts alike. The most prominent feature in this image is the bright central concentration, known as Sagittarius A, within which sits a cauldron of hot, swirling gas, imploding stars and exploding shells, all identifiable in images taken with progressively higher magnification.

Radio astronomers can magnify a patch of the sky, while preserving the eye-pleasing clarity, by looking for light with a smaller separation between the crests, that is, with a shorter wavelength, because for a given radio dish-size, the number of captured crests increases when more of them are squeezed into the array. Often, however, a magnified view displays new complexities in structure that may not always be apparent with the sliding scale of colors employed for the larger image. The number of chosen hues may simply be too small to adequately separate out the many new details that emerge. As such, the color coding is not always the same from figure to figure, even though some of the features may be in common.

Take the narrow blue filaments launched from the central splash of light in figure 1.5. These traverse the galactic plane, which runs from the upper left to the lower right. They trace the configuration of intense magnetic field activity in this region, and are

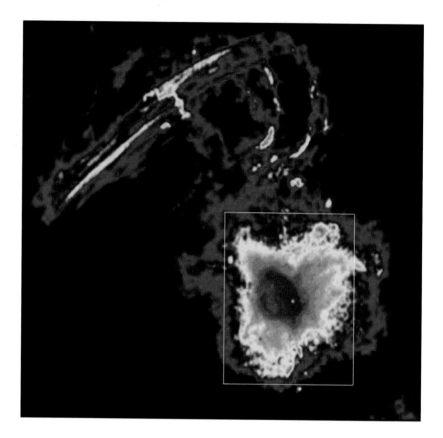

Figure 1.5 Magnified view of the central, bright region in figure 1.4. This image, produced by the Very Large Array, N.M., shows the intensity of radiation with a wavelength of 20 centimeters, produced mostly by magnetized, hot gas between the stars. The size of this magnified region is about 1/5 of that shown in figure 1.4, spanning a couple of hundred light-years in either direction. Again, the galactic plane runs from the top left in this image to the bottom right (as in figure 1.1). One of the most interesting features appearing here is the system of narrow filaments (some wrapped around each other) with a width of about 3 light-years. These radio filaments are oriented perpendicular to the galactic plane. The bright Rosetta-like region surrounds the center of our galaxy, which is magnified further in the following images. The spot in the middle of the red spiral identifies the radio source known as Sagittarius A*, believed to be the supermassive black hole at the center of our galaxy. The box to the lower right in this image shows the central 25 light-year region magnified in figure 1.6. (Image courtesy of F. Yusef-Zadeh, and the National Radio Astronomy Observatory/Associated Universities, Inc.)

evident also in the lower-magnification image of figure 1.4, identified by the yellowish-white streaks smudged across the middle of the photograph. These filaments, it turns out, are produced by very energetic particles flung around tight magnetic field lines, not unlike what happens when a penny is thrown into the strings of a harp. Others have suggested that these filaments may be the relics of galactic nucleation seeded by cosmic strings. Though intriguing, this hypothesis does not appear to be tenable since there is no natural way to link the properties of these speculative entities to the actual morphology of the observed threads. Any alternative theory that tries to explain them, however, must contend with the daunting task of accounting for their highly ordered structure over uncharacteristically long distances.

Figure 1.5 shows what the inner few hundred light-years of our galaxy would look like if our eyes were attuned to radio light with a wavelength of 20 centimeters. The central Rosetta-like structure reveals itself to be a composite of several individual elements, including the large, expanding shell from a powerful explosion that occurred at the galactic center sometime within the past 100,000 years. Although this hot bubble of gas looks in many ways like the remnant of a catastrophic collapse of a dying star after it has spent all of its nuclear fuel (i.e., a supernova), it is difficult to reconcile this interpretation with the energy required to produce this structure. It appears that as many as 50 to 100 supernovas would need to go off virtually simultaneously in order to create such a grand design. This special remnant, known as Sagittarius A East, is unique in our galaxy, and may instead have been produced by the explosive reaction of a star that ventured to close to the supermassive black hole, about which we will have much more to say later. Calculations show that these explosions ought to occur about once every 100,000 years or so. Like a comet approaching the Sun, rounding behind it and then receding to the far reaches of the solar system, the doomed star plummets toward the center and attempts to escape around the other side. Unlike the comet, however, it is compressed by the inexorable and unimaginably strong gravitational field of the black hole and explodes with the power of 50 supernovas to create the fireworks on display near the lower-right portion of this image.

The red source within the yellow remnant of the explosion is known as Sagittarius A West. We will have a clearer view of what it represents in the highly magnified images that follow (see, for example, figure 1.6). But here we acknowledge for the first time the appearance of a mysterious, pointlike object in the middle of the red and yellow fireball. Apparently, it defines the exact location of the Milky Way's center, and its radiative properties are unlike those of any other source in the galaxy. Its name is Sagittarius A*, and for reasons that will become clearer as we progress through this book, it is the deceptively unpretentious face of a 2.6 million solar-mass behemoth lurking in the middle of this inferno.

The exploding shell region of figure 1.5 is magnified further in figure 1.6, constructed from radio measurements at 6 centimeters. Using this shorter wavelength, we are now able to see features as small as a fraction of a light-year. The central red structure— Sagittarius A West—is barely 10 light-years across. We don't see the pointlike Sagittarius A* here since the chosen color scale saturates the radiative intensity in the middle in order to give us a vivid impression of the three-arm spiral. This hot gas is moving so rapidly in a counterclockwise rotation that photographs taken five to 10 years apart can actually reveal displacements of some features relative to each other, even at the 28,000 light-year distance to the galactic center (see figure 2.3). The velocity is as high as 1,000 kilometers per second in some places, indicating that a very strong gravitational force must be present to keep this gas from escaping the galaxy. In the late 1980s, the Nobel Laureate Charles Townes and his coworkers identified this motion, which was at that time inferred from the wavelength-shift of certain spectral lines, as an early piece of evidence for the presence of a large concentration of dark matter in the nucleus, which presumably provides the strong pull that keeps the gas motion in check.[2] We will return to this important discussion in chapter 2.

---

[2]Read a detailed report on this work by Serabyn et al. (1988).

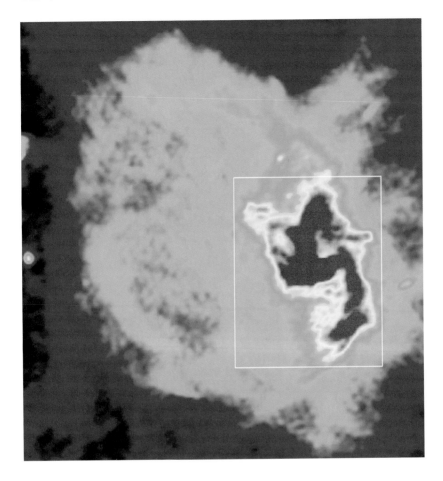

Figure 1.6 On a scale of about 25 light-years, the bright central region of figure 1.5 comprises two main components in this VLA radio continuum image (here rendered at a wavelength of 6 centimeters). The yellow shell-like structure (known as Sagittarius A East) is due primarily to particles radiating their energy as they spiral in a magnetic field. Its characteristics suggest an explosive origin, some 10,000 to 100,000 years ago, perhaps from the disruption of a star that ventured too close to the massive pointlike object in the nucleus. At the center of this shell, we see a spiral-shaped structure (mostly in red) of hot plasma radiating as it cools. The red feature, which orbits within 10 light-years of the central source of gravity, is known as Sagittarius A West. The weak extended blue region is a halo of thin gas surrounding this peculiar remnant. The box to the right in this image shows the central 10 light-year region magnified in figure 1.7. (Image courtesy of F. Yusef-Zadeh at Northwestern University, and the National Radio Astronomy Observatory)

1.3   THE PRINCIPAL CONSTITUENTS

We are now poised to plunge deeper into the center, to a distance as small as one-twentieth of a light-year from the location of our massive target. Compare this with the typical distance—a couple of light-years—between stars in the solar neighborhood, and you can begin to sense that we are in a territory where the physical conditions become exotic, if not extreme.

Shimmering and ethereal, the Sagittarius A West spiral (figure 1.7) turns harmoniously with the grandeur of a galactic whirlpool. Why three arms, we wonder? Are they really all connected? A peek ahead to figure 1.9 on page 22 reveals that at least one of these limbs—the one at lower right—is scraping the inner edge of a doughnut-shaped torus of molecular gas. We don't see it here because the radio telescope was tuned to the specific wavelength of the radiation produced by the spiral itself. We do have some evidence based on the velocity measurements of the gas shown in figure 2.3 that the vertical arm is actually a tongue of gas cascading downward toward the nucleus. These individual peculiarities notwithstanding, all three limbs are engaged in a beautiful ballet that sweeps the hot, ionized gas around Sagittarius A* like dancers encircling the royal throne.

As we magnify the view further (figure 1.8), we begin to see evidence for a dynamic interaction between the principal players on this stage. To the north of Sagittarius A*, one can just make out a cometary-like feature with a tail pointing away from the nucleus. This object is the envelope of a red giant star that is being stripped and blown away by a powerful wind originating somewhere near the center of the image. Infrared photographs of this region (see figure 1.12) reveal the presence of strong wind-emitting stars whose combined mass expulsion reaches levels of one solar mass every couple of centuries, which is sufficient to evaporate this gentle red giant and drive the efflux upward to high galactic latitudes.

The prominent character in this play, Sagittarius A*, is unmoved by the surrounding commotion. An inspection of its

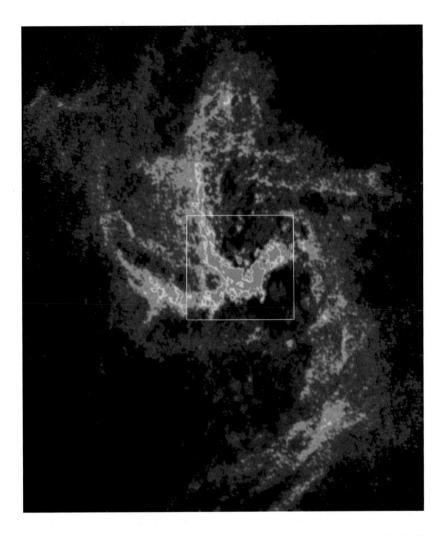

Figure 1.7 This false color image, representing a magnified view of the (red) spiral structure in figure 1.6, shows the 6-centimeter radio emission of highly ionized gas in orbit about the galactic center. Each of the "arms" is about 3 light-years in length, though it is not clear whether we are witnessing a real spiral pattern or merely a superposition of independent gas flows into the center. Nonetheless, astronomers have now determined that this gas is moving about the center with a velocity as high as 1,000 kilometers per second. The box in the middle of this image shows the central 2 light-year region magnified in figure 1.8. (Image courtesy of F. Yusef-Zadeh at Northwestern University, and the National Radio Astronomy Observatory)

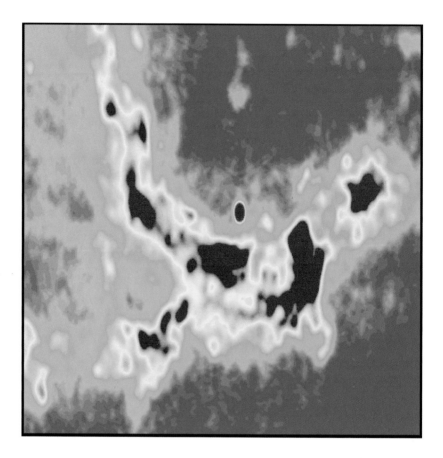

Figure 1.8 As we magnify the central (boxed) region of figure 1.7 further, we begin to see the structure of plasma radiating at a wavelength of 2 centimeters. The prominent features are the central portion of the spiral pattern of Sagittarius A West, and a bright pointlike source known as Sagittarius A*, near the middle of the photograph. We will learn that this object is associated with a mass of several million Suns and that it resides right at the center of our galaxy. We first noticed this spot of emission near the middle of the "Rosetta" in figure 1.5. To the north of Sagittarius A*, the cometary-like feature (in light blue against the dark blue background) is associated with a luminous red giant star that stands out near the top of figure 1.12. The gas blown upward from its envelope provides evidence of a strong wind emanating from the region near the supermassive black hole. The distance between Sagittarius A* and the red giant is approximately 3/4 of a light-year. (Image courtesy of F. Yusef-Zadeh at Northwestern University, and the National Radio Astronomy Observatory)

position shows that it has not budged relative to the projected location of distant quasars over several decades of continuous radio monitoring.[3] These bright beacons are so distant that for all intents and purposes their position is fixed to the firmament, providing us with a virtual grid to track the motion of nearby objects. And on this grid, Sagittarius A* has sat measurably still while the stars in its proximity (see figure 2.2) and the surrounding hot gas have swirled around it.

Sagittarius A* simply does not respond to its environment; it truly is the big gorilla that remains unfazed while all around it flail in frenzy. One can estimate how massive an object must be for it to respond—or, more accurately, to not respond—in this fashion, given the overall hustle and bustle at the galactic center. The most up-to-date calculations endow it with a mass of at least 1,000 Suns, which already rules out any known stellar object as the possible culprit underlying its behavior. As we will see in chapter 2, its mass is in fact significantly greater than this, though the argument based on its perceived lack of motion is an independent and important confirmation that its nature is nonstellar.

The now familiar spiral structure of Sagittarius A West provides a useful backdrop for the montage shown in figure 1.9. This image is unusual in that it demonstrates the interplay between two components that choose to reveal themselves with a light of very different color. By now we know about the three-arm spiral, whose gentle sweeping motion is anchored to Sagittarius A*. But when we look at the galactic center with sensors that detect far infrared radiation, we also see the manifestation of a molecular torus (rendered here in violet) orbiting the nucleus in a counterclockwise rotation. Its frail appearance belies how much mass it contains— at least 10,000 Suns' worth of molecular gas fills this ring. We see it at infrared wavelengths due to the warm dust trapped within it. Near Sagittarius A*, about a dozen very bright, blue stars (see

---

[3]This work can be rather challenging for radio astronomers, but the precision of their measurements improves with time. For a technical discussion on this topic, see the papers by Backer & Sramek (1999) and Reid et al. (1999).

figures 1.10 and 1.11) pump out 10 million Suns' worth of optical and ultraviolet radiation that is heavily absorbed by this dust, recalling the analogy with the water waves lapping against the gondola, producing a situation in which the heated particles glow with the infrared warmth of embers in a well-stoked camp fire.

Like cogs in the inner workings of an elaborate clock, the cartwheel and ring separately perform their exquisite movement on rigid axes linked to the common hub centered on Sagittarius A*. In the galactic center, this becomes a recurring theme. As we uncover each new dynamic entity, invariably we also discern its direct connection to this massive anchor, be it a fleeting wisp of thin gas, an orbiting star, or a large ordered structure threaded by magnetic fields.

At this point, we are not yet even close to our final destination, but let us pause briefly to collect our thoughts and consolidate our discoveries. We have explored the galactic center almost exclusively using radio waves, uncovering structure and activity for which not even the best optical images of the Sagittarius region could provide a precedent. The radiation spectrum is far more extensive than this, however, and it would be foolish for us to believe that we have exhausted our means of exploration. Space dust is a serious detriment to our imaging capabilities only for visible light, and we have not yet tapped into the possibility of looking at Sagittarius with an infrared telescope (other than our brief far-infrared view of the molecular torus in figure 1.9), or—which could prove equally fascinating—searched for the ripples of light at much shorter wavelengths, X-rays and gamma rays. Observations of these rays, which may bounce off the dust particles but are not otherwise seriously disrupted, are best made from the relative vacuum of space, and we will continue with our journey in the next section. For now, our travel log may be summarized schematically, as shown in figures 1.10 and 1.11.

These figures illustrate, in proper juxtaposition, the principal elements at play near the galactic center, especially the manner in which they relate to Sagittarius A*. The cluster of bright, blue stars to the left is not really isolated, but represents a concentration of such objects rendered more fully in an actual photograph

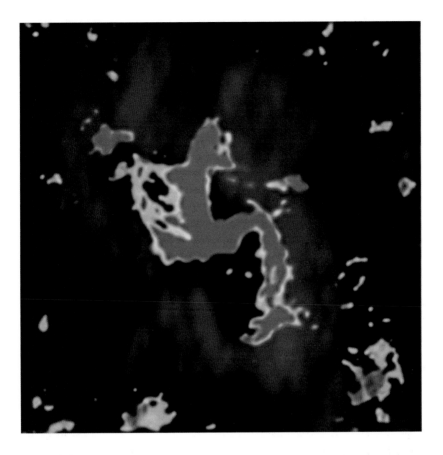

Figure 1.9 The images we have seen so far of the galactic center were taken at radio wavelengths. At optical wavelengths (see figure 1.1), the central region of our galaxy is mostly obscured. There is also some obscuration at infrared wavelengths, but not as severe as in the optical. This image, produced by superposing the radio photograph of the spiral in figure 1.7 with the view provided by millimeter cameras (shown in violet), provides evidence for the presence of a torus of dusty gas in orbit about the central source of gravity, Sagittarius A*. The dust in this ring shines by converting ultraviolet light into an infrared glow. The hot gas within it, however, radiates most of its energy at radio wavelengths, which are then converted into the optical colors recognized by the human eye. A more detailed description of the features seen here is provided in the schematic diagrams shown in figures 1.10 and 1.11. (Image courtesy of F. Yusef-Zadeh at Northwestern University, M. Wright at the Radio Astronomy Laboratory, UC Berkeley, and the National Radio Astronomy Observatory)

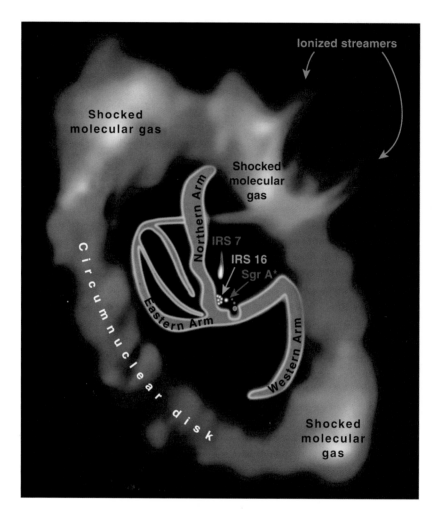

Figure 1.10 This cartoon of the inner 15 light-year region of our galaxy shows the exotic collection of astrophysical phenomena existing within the cavity of the circumnuclear disk (seen in figure 1.9). The dominant source of gravity is the strong radio source Sagittarius A*, which gives all indications of being a supermassive black hole. The nearby cluster of at least two dozen hot blue stars (known as IRS 16), bathes the entire cavity with ionizing radiation, and seems to be the source of a powerful wind that sweeps out from the central region with a velocity exceeding 700 kilometers per second. This wind is the apparent cause of the cometary tail associated with the red giant star IRS 7 in figure 1.8. The region surrounding Sagittarius A* is magnified in the view shown in figure 1.11. (Illustration by Linda K. Huff, reprinted by permission of the artist)

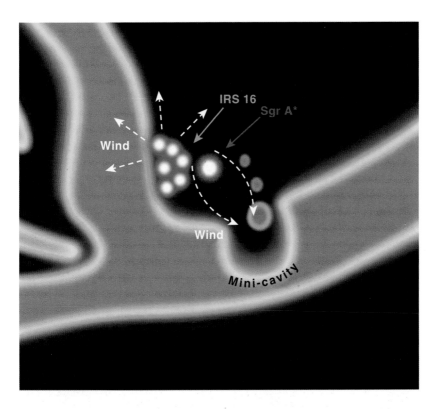

Figure 1.11 The existence of a peculiar hole (known as the "mini-cavity") in the radio-emitting gas surrounding Sagittarius A* may be due in part to the effects of winds emanating from the bright, blue stars to the left and focused gravitationally by the supermassive black hole, Sagittarius A*. (Illustration by Linda K. Huff, reprinted by permission of the artist)

of this region (figure 1.12). And to the right of our main character, there appears to be a mini-cavity in one of the spiral arms, which may have been produced in part by the impact of a concentrated wind flowing from the direction of Sagittarius A*. It is thought that the powerful wind from the bright, blue stars is focused gravitationally by the central massive object, producing a train of plasma blobs that help to carve out the hot, ionized gas as they plow into the orbiting material. There is no doubt that the galactic center is an intricate region, with a personality forged from

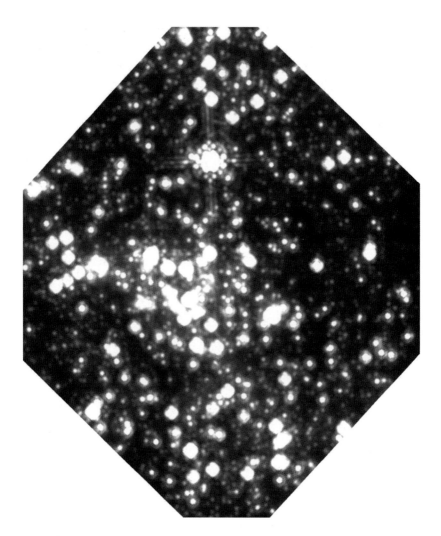

Figure 1.12 A near infrared photograph of the inner 2 light-year region of our galaxy, taken with the NICMOS 1.6-micron detector aboard the Hubble Space Telescope, shows the presence of a dense stellar cluster surrounding the nucleus. This field is the same as that shown in figure 1.7. The difference is that at this particular wavelength, most of the emission is from stars, whereas in the previous images we were witnessing the radiation produced by diffuse gas between the stars. Interestingly, the central source Sagittarius A* emits very little at this wavelength, so it does not even show up here. We will see a magnified view of the region surrounding Sagittarius A* in figure 2.1. (Image courtesy of M. Rieke and the NICMOS team at the University of Arizona)

the traits of its characters and their relative interactions. And yet, other than the bright, blue objects to the left of Sagittarius A*, we have hardly mentioned the role played by stars in the overall ensemble. Surely they must figure in somehow, given that much of what we know about the Milky Way is based on the light we see from them. Indeed they do, and in a rather surprising way. In the next section we will introduce these new cast members, and devote most of chapter 2 to a description of the breakthrough measurements that they have allowed us to make in recent years.

1.4   EXPLORATION FROM SPACE

In all the radio images we have examined thus far, other than the enigmatic, immovable object at the heart of the Milky Way, there are no points of light to be seen anywhere. There is a very simple reason for this—stars emit very little radiation at radio wavelengths. Evolution has served us well with our eyes, fine-tuning them to sense light with a wavelength corresponding to the Sun's dominant output, which we call "visible light" for obvious reasons. To survive optimally in our environment, we must have senses tuned to the medium conveying the most detailed information, and since Earth's atmosphere is bathed with the Sun's rays, it is sensible for us to be able to "see" this light. Most of the stars in the galactic center, like the Sun, should therefore be readily visible to our naked eye, though not to a radio telescope. Unfortunately the presence of space dust makes this otherwise simple task virtually impossible.

All is not lost, however, because it turns out that dust's deleterious effects weaken quickly as the separation of the light crests increases. In the canals of Venice, a water wave with a wavelength four times the size of the gondola is significantly less intrusive on the boat's stability than one with two. Well then, what if we could photograph the galactic center using an infrared camera, sampling the light produced with a wavelength longer than that to which our eyes are attuned, but short enough that the stars are still visible to that device?

Attaining this goal has been one of the greatest achievements
in our relentless campaign to uncover what is happening at the
galactic center, though not without its challenges. Stars twinkle
for the same reason that airplanes experience turbulence. Earth's
atmosphere is simply not static—it moves—and the light passing
through it is deflected, causing the star's intensity and apparent position to fluctuate. Neighboring stars, particularly those
huddled close to Sagittarius A*, are therefore not separable on a
photographic plate unless something is done to stop the twinkling.
The advent of the space age toward the end of the twentieth century has made it possible for us to avoid this problem altogether
by moving our platform above the atmosphere. In this regard,
the deployment of the Hubble Space Telescope in orbit around
the Earth will be remembered as one of the most notable scientific advances ever. Its array of instruments included a 1.6-micron
camera with the capability of taking infrared photographs of the
heavens with a clarity never before approached in human history.

Peering toward the galactic center, the Hubble Space Telescope
recorded the image shown in figure 1.12, which finally reveals to us
the dense stellar cluster compressed within the inner 2 light-years
of our galaxy, the same region we viewed at radio wavelengths in
figure 1.8. Recall that in the solar neighborhood, there would be
just a single star within this same volume! Sagittarius A* lurks
specter-like in the very middle of this picture, but it does not emit
infrared light so we have no way of studying it *directly* with the
Hubble Space Telescope. However, its overpowering influence on
the stars around it is enormous, and much of chapter 2 will be
dedicated to this phenomenon, because it is currently the best
means we have of measuring Sagittarius A*'s mass. We recognize
in this photograph the concentration of bright stars to the left of
where Sagittarius A* would appear, constituting the blue stellar
cluster we identified earlier in figures 1.10 and 1.11. We also note
to the north the appearance of the very bright, red giant, whose
envelope is evaporating to form the cometary blaze we saw in
figure 1.8.

In chapter 2 we will see a magnified view of the central region
of this infrared photograph, and learn that some of these stars

dart across the sky with the highest velocities in the galaxy (some as fast as several thousand kilometers per second), venturing reck- lessly to within a mere twentieth of a light-year from Sagittarius A*, before engaging in a hasty retreat. It is possible that even smaller stars, too faint to be seen in this photograph, though far more numerous than the bright ones captured in this image, may steal even closer and, occasionally, fall in. One of them may have produced the huge explosion whose remnant signature is evident in the glorious fireworks on display in figure 1.5.

The advent of the space age has also given us the means to view the heavens at very high radiation energies, where the effects of space dust are equally insignificant. X-rays and gamma rays traversing Earth's atmosphere are readily absorbed by the atoms within it, which is fortunate for us as organic beings since these high-energy rays would otherwise quickly kill living cells and ren- der Earth incapable of supporting life. But out in space, where the density of gas and dust are much lower, X-rays and gamma rays flow freely over very long distances. During its time in orbit, the Compton Gamma-ray Observatory did for gamma-ray astronomy what the Hubble Space Telescope has been doing at infrared and optical wavelengths. Before its fiery descent, plunging into the Pacific Ocean during the summer of 2000, this bus-sized detector enabled us to answer the question: "What would the sky look like if we could see gamma rays?"

The answer is figure 1.13, in which the firmament seems to be filled with a shimmering high-energy glow diffusing throughout the galactic plane. In the last decade of the twentieth century, Comp- ton produced several spectacular images of the sky, sensing radi- ation with more than 40 million times the energy of visible light. Like many of the other images shown in this chapter, the colors we see in this figure are coded to represent a range of intensities our eyes would detect if they could see that particular type of ra- diation (in this case, gamma rays). Some of this light is produced by the most exotic and mysterious objects in the universe. Within the galactic plane, much of the high-energy radiation is produced by pulsars—rapidly spinning, magnetized neutron stars barely as

Figure 1.13 High-energy-sensing instruments have difficulty resolving features as small as those seen in the previous radio images, since each photon entering the detector carries a greater punch, and is therefore more difficult to corral. Nonetheless, in the course of a full-sky survey, the EGRET instrument aboard the Compton Gamma-Ray Observatory detected a very-high-energy source that is coincident with the galactic center. This false-color image, showing the gamma-ray intensity within a field of 10 degrees by 30 degrees (so roughly half of the image-size seen in figure 1.1) is color-coded such that white represents a count rate 300 times higher than that of black. In this view (for which one degree corresponds roughly to 430 light-years) the galactic plane (shown as a yellow line in figure 1.1) lies in the horizontal direction. The photon energy sampled in this image is greater than 1 billion electron volts. (By comparison, the electron in a hydrogen atom is stripped away from the nucleus with an energy of only 13.6 electron volts.) Given that the angular resolution of the gamma-ray detector is relatively poor, it is not clear from just this particular observation alone whether the gamma-ray source at the galactic center is associated with a pointlike object, or whether it is diffuse over a region of up to 300 light-years in extent. But recent theoretical work suggests that the gamma rays are produced by the remnant of a hyper-explosion (shown in figure 1.6) that occurred at the galactic center sometime within the past 100,000 years. (Image courtesy of J. R. Mattox in the Department of Physics and Astronomy, Francis Marion University, and NASA)

big as a city, though formed in the violent crucibles of stellar explosions—stellar-sized black holes, such as the well-known Cygnus X-1, and hyper-energetic hydrogen nuclei, known as cosmic rays, which can easily puncture the exterior walls of a spacecraft. Above and below the plane, quasars beyond our galaxy are powered by supermassive black holes like the one at the galactic center, and produce gamma-ray beacons out to the edge of the visible universe. However, the most prominent source for the high-energy radiation is clearly the galactic center, which illuminates the inner 2,000 light-years of the Milky Way. Curiously, the centroid of this intensity is not Sagittarius A*, as many may have expected, given that black holes can be prodigious sources of power, but rather the explosive remnant Sagittarius A East (figure 1.5). In retrospect, that should not surprise us since the process that created this fiery corona appears to have been the most energetic event of the past 100,000 years.

The Compton Gamma-Ray Observatory was also designed to have the remarkable capability of detecting the radiation produced when matter, in the form of electrons, annihilates with antimatter, in the form of positrons. Given that the mass of these particles is always rigidly fixed, the light they produce when they collide escapes with the finely tuned energy of 511,000 electron volts, which can be distinguished from the rest of the radiation field. The fruits of this engineering and scientific feat produced one of the most memorable images ever seen of the sky, shown with the galactic plane oriented horizontally in figure 1.14. We are witnessing here the annihilation of matter and antimatter on a grand scale, stretched across a 20,000 light-year portion of the galactic plane.

Jules Verne's Professor Hardwigg would quickly point out that notwithstanding the immensity of this panorama, the action is clearly focused on the nucleus. Indeed, to produce such an eerie specter, some 10 million billion billion billion billion (10^{43}) positrons must be sacrificed each and every second at that location. What we do not yet know for sure is whether this antimatter is produced by Sagittarius A*, or one or more members of its entourage. This uncertainty arises because high-energy gamma rays pack an abnormally strong punch and are difficult to corral, so although the Compton Gamma-Ray Observatory could easily count how many were reaching it from a given direction, it could not with precision determine their exact origin. In addition, the glow we

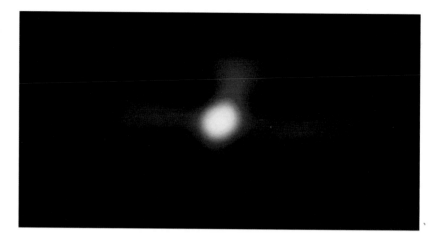

Figure 1.14 This map, spanning a region approximately 20,000 by 10,000 light-years, was produced by the OSSE instrument aboard the Compton Gamma-Ray Observatory. It reveals where antimatter (in the form of positrons) is being annihilated near the galactic center. When positrons annihilate with electrons, they produce characteristic photons with an energy of 511,000 electron volts, whose intensity is shown as a function of position in this color-coded map. The map shows evidence for three distinct features: (1) a central bulge, (2) emission in the galactic plane (which lies in the horizontal direction in this image), and (3) an enhancement or extension of emission above the plane. (Image courtesy of J. Kurfess at the Naval Research Laboratory, and W. Purcell in the Department of Physics and Astronomy, Northwestern University, and NASA)

see in figure 1.14 is definitely diffuse, rather than pointlike, so the positrons live a long time before they meet their partner and perish, which allows them to wander far from their originating site. Future high-energy space missions will have a significantly better spatial resolution, which will allow us to complete our investigation of this phenomenon and finally point to the true source of antimatter in the galaxy.

And so we have begun developing a case for the unusual character of Sagittarius A*. In the next chapter, we will complete our inward trek and assemble the evidence that compels us to accept as real its supermassive black hole status. The implications for our view of how the universe functions are profound, touching on the most enigmatic prediction of general relativity—the existence of separate worlds largely disconnected from the one in which we live. In chapter 3, we will develop this theory from first principles and reach an understanding that will facilitate our interpretation of the observational vistas we are uncovering during this journey to the center of our galaxy.

2

CONDENSATION OF DARK MATTER

A mind-numbing 1 trillion Suns fill the whirlwind of dazzle and color that is the Milky Way, and yet over most of the galaxy, stars account for but an infinitesimal fraction of its volume. If we were to shrink these galactic wanderers to the size of a cherry, we would need to commute the distance between the major cities in Europe to simulate an excursion from one star to the next. Such is the empty expanse between the Sun and its nearest neighbors. Twenty-eight thousand light-years away, however, some 10 million stars swarm within a mere light-year of the nucleus—that's like placing 10 million cherries within a diameter of just 1 kilometer. The brightest members of this crowded field are captured in figure 2.1, a magnification of the view near the middle of figure 1.12. But don't be fooled by this photograph's deceptive calm—many of the stars located close to Sagittarius A*, indicated by the cross mark, are orbiting at blistering speeds of up to 5 million kilometers per hour, allowing us to see their motion in real time. In this chapter, we will be using this information to not only "weigh" the dark matter content of the nucleus, but perhaps more glamorously, also confine it to within a small volume of space no bigger than the inner planetary orbits of our solar system.

The galactic center's accessibility in this regard is the chief reason why its dark matter concentration has become the prime focus of black-hole research, for nowhere else in the universe can astronomers study the behavior of objects this close to the nucleus.

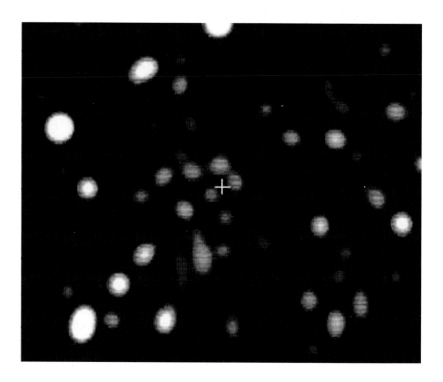

Figure 2.1 Using state-of-the-art shift-and-add speckle image reconstruction, Menten and his collaborators produced this 0.00022-centimeter image spanning the inner 0.2 by 0.2 light-year region of the central star cluster surrounding the location (marked with a cross) of Sagittarius A*. The size of the cross itself here reflects the uncertainty in the positioning of the radio image relative to the infrared localization. This wavelength samples the emission of red giant stars, which show up clearly in this photograph. The fact that Sagittarius A* is not visible at this wavelength, within this instrument's sensitivity, implies that it is predominantly a radio source, which places rather severe constraints on any model attempting to account for its radiative characteristics. Some of the stars closest to Sagittarius A* have been seen to move with a velocity of several thousand kilometers per second (see figure 2.2), which confirms, through Kepler's orbital law, the "measured" mass of almost 3 million Suns at the center. These stars move so fast, in fact, that photographs taken just a few years apart reveal their accelerated trajectories about the nucleus. (Image courtesy of K. M. Menten et al. at the Max Planck Institut für Radioastronomie, Bonn, Germany)

2.1   A SWARM OF STARS

The challenge of identifying individual stars in this confusion of motion and light at the distance to the galactic center may be met either by moving the telescope above Earth's atmosphere (which produced the photograph in figure 1.12), or by employing a rather clever technique for removing stellar twinkle from the ground. The problem, to which we alluded in chapter 1, is that the air moves unevenly, distorting the stellar light fleetingly, though significantly, so that a long exposure of the sky produces a blurred image. Those of us who have attempted to take action shots of athletes in motion will recognize this difficulty with empathy. The technique developed to counter the annoying defocusing is rather simple, as it turns out; it requires taking thousands of very quick, high-resolution photographs instead of a single long exposure. Although each snapshot in this sequence is somewhat faint, an overall bright image can still be produced if these photographs are added together, stacked one atop the other, but displaced ever so slightly from each other in just the right amount to compensate for the movement of the image induced by the atmospheric distortions.[4] The resolution achievable with this technique, using for example the 10-meter Keck I Telescope—the world's largest infrared telescope—on Mauna Kea in Hawaii, is equivalent to seeing two cherries 10 feet apart in Paris from the top of the Empire State Building in New York.

We will learn that with the incredibly high density of objects in such a tight confinement, the matter content at the galactic center is far greater than we imagine based on the stars we actually see using this so-called shift-and-add technique. The term "dark" matter has come to symbolize the source of strong gravity we often measure in certain pockets of the universe, where the meager

---

[4]This effort has benefited from the contributions made by several observatories around the world. The principal investigators leading the effort to use these techniques for imaging the galactic center have been Andrea Ghez, Mark Morris, Eric Becklin and their collaborators at UCLA, and Reinhard Genzel, Andreas Eckart (now at the University of Köln), and their collaborators at the Max Planck Institut in Garching, Germany.

stellar content—the "visible" matter—does not appear to be up to the task of providing the necessary pull. It will become apparent that the visible matter near Sagittarius A* can provide at most 1 percent of the required attraction to keep this volatile region in harness, implying that an overpowering dark component, or components, must be lurking nearby. And yet, the central stellar density is still far greater than anywhere else in the galaxy, so it is natural to wonder how these stars arrived at their captive endpoint.

Stars can live a long time, but they do not always hide their age well. Their color and their mass reflect rather accurately the span of time that has passed since their nascence. In the galactic center we seem to be endowed with at least two generations: one that formed some 10 million years ago, basically a wink of an eye on a cosmic time scale, and another whose birthrate peaked about 100 million years earlier. The prevailing thought on what can cause this generational divide is that there have been multiple epochs of star formation near Sagittarius A* possibly triggered by the sporadic infall of fresh material toward the center. In other words, new stars form whenever a fresh supply of new gas is swept into the central region by the strong gravitational pull of the nucleus. One is almost tempted to invoke a seasonal analogy, in which spring rains bring heavy growth.

But stars eventually die, becoming compact cooling embers that fade with time (see chapter 3), so generations of withered Suns must be floating among the living. They are no doubt there in numbers, and we have circumstantial evidence that at least some of the dark matter surrounding Sagittarius A* is in the form of these ashen cores. For awhile, it was even thought that the great ponderous character of Sagittarius A* itself was nothing more than these huddled dying masses—to be sure, a rather morbid vision, but an altogether comprehensible one. In the next section we will see why this is no longer tenable, through the skillful tracking of impudent stars that have ventured to within hundredths of a light-year of the middle. There very few of their dead brethren have settled, and the influence of these dead stars must therefore be minute.

2.2   THE MOTION OF STARS AND GAS

We marvel at the numbers and wonder what they mean, but our analytical mind relies on motion and change to divine how things work. In the end it is the dynamics of the galaxy's central constituents that unfold the mystery. We could hardly care about Sagittarius A*'s entourage if all they did was sit and wait. As it happens, they are reckless, and move with frenzy.

To gain a concrete sense of how we can use this motion to learn about the forces at play in the central region, consider a very simple experiment whose only device is a rock connected to a long string. By twirling this overhead, you will demonstrate in easy fashion the essential behavioral elements of stars locked in orbit around Sagittarius A*. What you should notice first is that unless the rock is moving with some minimum speed, it does not execute circular motion; you can start the process, but the rock quickly falls. Above the minimum speed, however, the rock can circle comfortably, but the strength with which you must hold onto the string is greater, the faster the rock moves. If you could twirl the rock fast enough, eventually the string would break. The effort you expend with your hand in holding the string is the gravitational force pulling the star inward, and the string is in a sense the indicator (a "vector") showing the direction to the source of attraction or pull, in this case your hand.

We will learn in chapter 3 that an object's gravitational force scales in proportion with its mass. The velocity of a star near the nucleus may therefore be used to measure the dark matter content of that region, since the faster the star moves, the greater must be the force of gravity preventing its escape, and therefore the greater must be the central mass. Moreover—and this is what makes the recent measurements described below so breathtaking—the matter holding the stellar orbits intact must clearly be contained entirely within those orbits, just as the hand stays inside the rock's path to keep it in circular motion. Experts and nonexperts alike recognize that it is one thing to measure the matter content of the galactic center, but it is an entirely new and significant leap forward to localize that mass within the tiny orbits now being tracked with high-precision imagery.

The efforts to track the motion of the 200 brightest stars in figure 2.1 with the Keck telescope in Hawaii and the New Technology Telescope in Chile have revealed that the 20 or so closest neighbors of Sagittarius A* exhibit velocities of a magnitude not seen—by far!—anywhere else in the galaxy. The staggering rapidity with which changes occur in this crowded field may be gauged anecdotally by the star that was closest to Sagittarius A* in 1995 and subsequently disappeared from view. Several possible theories have been proposed to explain this vanishing act. One possibility is that the stellar image seen in 1995 may have been a gravitational lensing event, which occurs when the light path from a star passing behind the massive object is bent by its strong gravitational field, a phenomenon we will carefully analyze in chapter 3. Another possibility is that the stellar image may have actually been a flare due to a star falling into Sagittarius A*. We may never know the truth.

Three other stars, known unceremoniously as SO-1, SO-2, and SO-4 (see figure 2.2), orbit a mere two-hundredths of a light-year from Sagittarius A*, and are zipping along so fast, at up to 5 million kilometers per hour (as we said before), that they move considerably on photographic plates taken only several years apart. They advance so rapidly, in fact, that it is now possible to unambiguously trace their orbits with startling precision, revealing periods as short as 15 years. Compare this with the 220 million years it takes the Sun to orbit the galactic center. Like the twirled rock constrained to move in a tight circle by the string pivoting at the hand, these stars exhibit paths whose "string" is pointing to the source of gravity causing their confinement, and all three of their accusatory vectors are directed toward Sagittarius A*. The mass required to harness their motion is *2.6 million Suns*, compacted to an excruciating density in a region much smaller than our solar system. Indeed, the best current estimates for the size of Sagittarius A* permit it to be no bigger than the orbit of Mars. We remind ourselves that at Earth's location, the same volume of space contains exactly *one* Sun. Squeezing dark matter to such

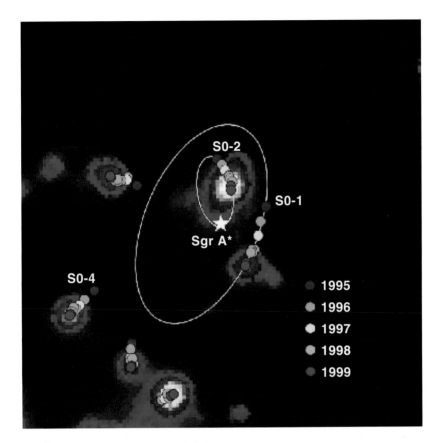

Figure 2.2 This image shows the orbits of two stars (labeled S0-1 and S0-2) around Sagittarius A* as inferred from the detection of acceleration in their proper motion over several years of observations. The image spans a 0.1 by 0.1 light-year region surrounding Sagittarius A*, and the individual positions of the stars at the various epochs are shown as colored dots. The underlying image was produced from observations of the galactic center at 0.00022 centimeters using the Keck telescope in Hawaii. (Image courtesy of A. M. Ghez, M. Morris, E. E. Becklin, A. Tanner, and T. Kremenek at UCLA)

a high density, without causing it to condense to a point, would have consequences that are excluded by the observations; we will therefore argue that this mass should be attributed to a single object, one that we will recognize as being a supermassive "black hole" in chapter 3. But we're getting ahead of ourselves, for this case is not yet shut. What about those dead stars, not to mention the active ones we see in images such as figure 2.1? Could they not mimic the effects of strong gravity?

Historically, it was the motion of gas in orbit around Sagittarius A* that hinted at its true nature, given that hot plasma may be identified easily using large radio telescopes whose spatial resolution can produce the glorious images we saw earlier. The excitement generated by the rapidly moving stars is far more recent, rendered possible only after the evolution in infrared astronomy took away the stellar twinkle. Gas is not immune to the crushing force of gravity when matter is abundant and, absent any other influence, it ought to display a behavior concordant with that of stars. However, there are some noteworthy differences, including the fact that stars orbit randomly in a frenzy around the galactic center, whereas the ionized gas is part of a coherent flow with a systematic motion that is decoupled from the stellar orbits. The majestic cartwheel motions we saw in chapter 1 would not otherwise be possible.

It is curious, though, that gas is there at all. We might expect that once it descends into the maw of Sagittarius A*, any trace of it should vanish, given how overwhelming the pull of 2.6 million Suns should be. By association, we might also wonder where the "rain" that fosters the new growth of stars originates. An answer may have been provided recently by the discovery of ammonia, in a mixture with other compounds, cascading out of giant clouds encircling the galactic center toward the torus of material surrounding the nuclear region (see figure 1.9). Located some 25 to 50 light-years from Sagittarius A*, these giant molecular clouds shed their outer layers, which eventually wind their way along narrow cones into the torus. The trickle apparently continues when gas and dust are stripped from the inner lining of the giant doughnut by the strong gravitational pull of the condensed dark matter, and these then spiral inward toward the nucleus.

A swoosh of hot wispy streamers is all we are likely to see, since these filaments thread the stellar cluster with speeds of up to 1,000 kilometers per second (see figure 2.3). Though these velocities are

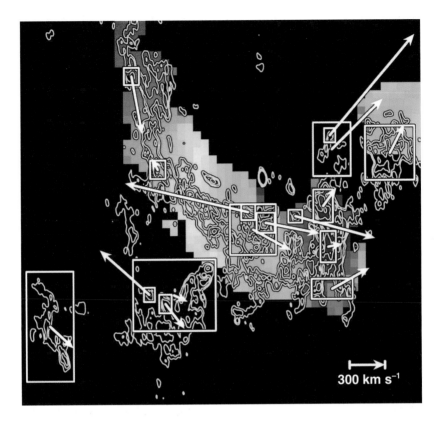

Figure 2.3 High velocities around the central source of gravity are not limited to stars. Gas too can be seen to flow with speeds not seen anywhere else in the galaxy. This image, based on proper motion measurements carried out with the Very Large Array radio telescope over nine years, shows the direction of ionized gas flow superposed onto a radio continuum image at a wavelength of 2 centimeters. The size of the region shown here corresponds to that of figure 1.8, but the color-coding now represents the gas velocity: on a sliding scale of colors, red indicates zero velocity and blue represents a value of 400 kilometers per second. The boxes are regions where proper motion measurements have been averaged; the velocities projected onto the plane of the sky are represented by white arrows. The bright circular feature close to the center of the image is Sagittarius A*. As is the case with the stars in figure 2.1, these gaseous features slide so fast that photographs taken several years apart can pick up the changes in the image brought about by their motion. (Image courtesy of F. Yusef-Zadeh and D. Roberts at Northwestern University)

no match for those of the frenzied stars, the gas streaks are much farther out, so using the twirling rock and string analogy, we infer a central gravitational tug for them of about the same magnitude as for the stars. Actually, the early assessment of the gas motions pointed to a slightly larger mass (of about 3 million Suns), but this is tempered somewhat by the realization that unlike the heavy stars, the plasma is thin and light and easily pushed aside by other influences, such as the powerful winds from the bright, blue stars to the left of Sagittarius A*.

That there lurks a dark concentration of matter at the location of Sagittarius A* is no longer in doubt. The next step is to identify the culprit (or culprits) and to uncover what beats at the heart of the Milky Way with a gravitational intensity not seen anywhere else in the galaxy.

2.3   THE MISSING MASS

Our journey has taken us deep into the forbidding, almost unscalable, recesses of the galaxy's inner sanctum, where even massive Suns find themselves subdued by a force that literally threatens to tear them apart—and occasionally does. But could not this gravitational pull simply be the concerted effort of all those stars we see in figure 1.12? Some forces in nature can cause particles to either attract or repel each other, depending on whether they have similar or opposite attributes, such as the sign of their electric charge. By cleverly arranging these particles in type and position, we could in principle cause a net cancellation of the force between them by balancing the attractions and repulsions. Gravity, however, always acts in the same sense for all matter, meaning that if the swarm of stars surrounding Sagittarius A* is made thicker, the resulting force pulling other objects inward would increase as well; one object's gravitational attraction cannot be canceled by the (nonexistent) gravitational repulsion of another. In principle, therefore, all of the effects we witness in the central realm could be induced by a tight clustering of inert stars pulling everything toward them in the middle.

This question is of such import that astronomers have set themselves the task of actually cataloging all the stars swarming in Sagittarius A*'s neighborhood. Under the right conditions, it is even possible to look at the rainbow of colors produced when the light from each star is passed through an appropriate prism or grating, which is a powerful technique given that variations in stellar mass produce an assortment of different colors. When combined with other clues, such as their individual brightness, it is quite straightforward to accurately infer from this their mass, and thereby chart the distribution of visible stellar matter in the nucleus. The situation in practice is slightly more complicated than this since most of the stars don't actually stand out as individual points of light like those in figure 2.1. The mass for them is instead inferred from their aggregate brightness, which completes the total assessment of how much matter is associated with the light we see from the galactic center.

The end result of this laborious effort has produced the yellow curve rising with distance from Sagittarius A* in figure 2.4. The precipitous plunge in luminous matter as we approach the galactic center (to the left) demonstrates dramatically why the potent gravitational pull of Sagittarius A* cannot be due to the stars we see. We know this because the motion of SO-1 and SO-2 and their brethren, through the twirling rock and string analogy, compels us to accept that the enclosed mass (white circles) levels off, approaching a *constant* value toward the origin. The divergence between the measured mass near Sagittarius A* and that of the luminous stars is hardly subtle. So where is the missing mass?

Other than attributing this to a single object at the location of Sagittarius A*, two other possibilities have presented themselves, though both appear to be ruled out now by the most recent measurements described earlier in this chapter. One of these proposes that the gravitational pull is due to a new type of condensed matter, made up entirely of ghostly particles known as neutrinos. But we will see in the next section that these particles cannot be the agents of subjugation for the commotion at the galactic center.

Almost certainly, some of the hidden mass must reside in the stellar graveyard collapsing toward the nucleus. With generation

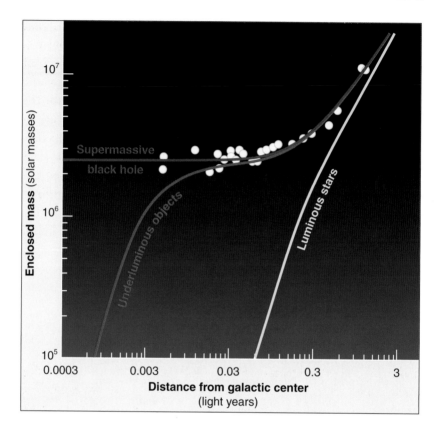

Figure 2.4 Using the kinematic information gleaned from the motion of stars and gas in orbit about the galactic center (see figures 2.2 and 2.3), we can now accurately plot the mass enclosed within a given radius all the way down to an unprecedented 0.045 light-year. The various data points in this graph represent different classes of objects. From 0.045 to 0.3 light-year, the enclosed mass appears to be constant with a value of 2.6 million Suns. By comparison, the enclosed mass contributed by the *visible* stars (i.e., the luminous matter seen in figures 1.12 and 2.1) is given by the yellow curve to the right. Alternative interpretations of this mass-radius relation, based on the idea that the dark central matter is in the form of underluminous, dead stars, are shown as a pink line. This fit is optimized to show as much flattening of the mass function at small distances as possible. None of these is an adequate fit to the data. The unprecedented high density of dark matter at the location of Sagittarius A*, which exceeds 100 billion Suns per cubic light-year, is indicative of a single pointlike object—a supermassive black hole. (This graph was produced using data supplied by A. Ghez, M. Morris, E. E. Becklin, A. Tanner, and T. Kremenek at UCLA.)

upon generation of stellar lives terminating in ashen cores that fade with time as they thrash about inside the gravitational pit, there is no doubt that the interstices between the points of light in figure 1.12 contain many dark stars. We know they're there because as we approach the nucleus, space gets heavier than the luminous matter. The pink line in figure 2.4, tracing the distribution of dark mass, deviates more and more as we venture inward and cross 1 light-year from the center. But just as the luminous matter fades to nothing near Sagittarius A*, so too does the underluminous mass charted here in pink. We will see that in light of the information now available, there are actually two fundamental reasons why the mass of Sagittarius A* cannot be due to these dark stars either.

2.4 A SUPER-HEAVY CENTRAL OBJECT

We are now in a position to look carefully at what might be a plausible option—other than a supermassive black hole—to explain the central 2.6 million Suns' worth of dark matter: that this mass is due to a clump of dark objects, stars that are either too low in mass to ignite nuclear reactions and shine like the Sun, or that have died (for example in supernova explosions) and are now cooling embers fading from sight. The latest measurements have given us two powerful reasons to argue that this alternative is no longer tenable. The first is quite evident from the shape of the pink curve in figure 2.4, which shows how dark matter in the form of underluminous stars distributes itself around the center. It now deviates from the measured total mass (white circles) by an amount that cannot be attributed to error or some glaring oversight, and this divergence gets bigger the closer we approach Sagittarius A*.

Let us see why this happens. Recalling once again the twirling rock and string analogy, we should feel comfortable with the idea that any object in the vicinity of this powerful gravitational pull cannot for long survive in that environment unless it is moving very fast. Like the rock, it will simply fall to the middle if it does not attain a speed above some minimum value. Dark stars are subject to the same laws of physics as the bright, active ones, and they too must be swift or plunge into the deep pit. Thus, if nature is to simulate a powerful gravitational field by compressing many dark stars into a tiny volume, the resulting swarm is literally abuzz with so many rapidly moving objects that the region within which they move has a finite size set by the range of speeds they possess. This size, it turns out, is somewhere between one-half and 1 light-year, *10 times bigger* than the orbits of stars such as SO-1 and SO-2 in figure 2.2. The pink curve in figure 2.4 shows us the best that nature can do with dark matter distributed in this fashion, optimizing the central condensation as much as possible at the expense of spreading the rest of the dark objects farther out. In other words, even if we could contemplate such a heavy graveyard, its residents would likely be spread out more than is indicated by the pink curve and their numbers would plunge even sooner.

The fact that the measured mass, which is indicated by the orange curve in this figure, does not change with distance within the inner two-tenths of a light-year from Sagittarius A* is a dead giveaway that nothing of significance can be happening to augment the gravitational pull within that region. The only explanation is that most of the dark matter is contained at a single point.

The second reason why a graveyard of dead stars cannot adequately represent the strong gravitational pull of Sagittarius A* is that in order to fit all of them within the orbits of SO-1 and SO-2, we would need to create a highly unstable, short-lived situation not suited to the apparent longevity of the galactic center. What we mean by this is that if by some miraculous reason all the dead stars could be squeezed by a potent force into such a tiny volume, it would take fewer than 100 million years for this assemblage to get completely disrupted. Think of piling up a large number of ping-pong balls on a flat surface. You may be able in this fashion to build a mound, but if you make it too steep, it will quickly spread, the balls will start colliding with each other, and the mound will disintegrate as they squirt in all directions. Concentrating a large number of dark stars very close to the center is like making a steep pile of ping-pong balls, so a highly uncommon situation such as this is not only difficult to arrange, but even more difficult to preserve. A period of 100 million years is far shorter than the age of the galaxy, which is believed to have formed about 12 billion years ago.

The other, more exotic possibility is that the enormous central mass is due entirely to a hypothetical giant ball of neutrinos, particles that congregate together gravitationally, but are otherwise very difficult to detect. But neutrinos don't condense that much, and a ball of 2.6 million Suns' worth of these little specters is expected to have a radius of no less than six-hundredths of a light-year. Two of the stars shown with orbits in figure 2.2 are well within this radius, only two-hundredths of a light-year from Sagittarius A*, so most of the concentration of neutrino mass would not be felt inside of their orbits. With a highly contrived distribution, it might still be barely possible to account for the stellar motion, though it now appears more and more that the existence of the particular species of neutrinos (known as sterile neutrinos because they interact with virtually no other particles in nature) required to make this work is unsupported by any other branch of physics. Explaining the dark matter at the galactic center with this scenario would also leave us perplexed by a situation in which some galactic cores have these condensations of unnecessary particles, while others do not, since a comparable ball of condensed sterile neutrinos cannot account for the dark matter in the nuclei of other galaxies (see chapter 6).

As far as our current knowledge can guide us, we have no choice but to accept the fact that most of the dark matter at the galactic center is contained within a single object, known as Sagittarius A*. It is quite remarkable that we have reached the point in our technological development where we can also localize its mass to a region no bigger than the inner portion of our solar system, a feat comparable to being able to tell which of two neighboring fruit vendors has the greater number of oranges to sell in Melbourne's Queen Victoria market from a vantage point in San Francisco.

We have become aware of the dark matter, we know where it resides, and we know that it has mammoth proportions. Sagittarius A* is a giant, ponderous and slow, and unfazed by the hubbub surrounding it. It weighs in at a portly 2.6 million Suns, yet has a girth no bigger than Mars's orbit. Our everyday experience cannot possibly prepare us for what lies in wait when one day we venture across the galaxy to behold its central realm. How does one even deal with the nature of space and time in a region where gravity is so strong that it threatens to tear reality apart? Our task is to make sense of what we've found and to retrieve whatever morsels the universe is throwing our way to uncover its function and its ultimate destiny. But to do that, we must first understand why it even matters that such a large, concentrated mass exists, and how it produces its inexorable, strong gravity.

3

THE THEORY OF GRAVITY

The question "Why does matter have mass?" may seem completely redundant, for we often take it for granted that we have a commonsense understanding of what mass and weight must be. Some may say that our everyday experience makes these concepts self-evident. In fact, they are not, and therein lies one of the main reasons for our struggle, since the time of Aristotle, to come to grips with what gravity is and why objects behave the way they do in the presence of other matter. We cannot even hope to fathom what it must be like near an object as big as the Sun, let alone one that is millions of times "more massive," until we understand the force—gravity—that gives them such a powerful influence on their environment. (Incidentally, the concept of gravity as a "force" is itself subject to review when we consider the general theory of relativity.) In this chapter, we will examine the evolution of thought that has brought us to a rather comprehensive (yet still incomplete!) theory, whose main prediction is the existence of objects that have collapsed completely unto themselves. We are apparently dealing with one of these at the galactic center.

3.1 WHAT IS MASS?

By now all of us have an intuition that gravity is associated with mass: in simple terms, gravity is an attractive force that any material object exerts on any other. However, we must be very precise about what we mean by this, for our experience with common phenomena makes us aware of mass in two quite distinct ways.

We all remember from high school that matter is made up of a very large number of smaller constituents called atoms, or on a slightly larger scale, combinations of atoms in the form of molecules. Water, for example, is the compound H_2O, meaning that two hydrogen atoms (H) combine with one oxygen atom (O) to form a molecule of this substance. Atoms themselves are not the fundamental building blocks, for they are assembled from more basic components called protons, neutrons, and electrons. When we "weigh" these particles using an appropriate balance, we find that the protons and neutrons are almost 2,000 times heavier than electrons. They occupy the central region of the atom (the nucleus), with the lighter electrons swirling around them. Several forces act to keep this system intact, including the "strong" force that binds the neutrons and protons together, and the electric force of attraction between the positively charged protons and the negatively charged electrons. By the way, the actual definition of what is positive and negative in this context is simply a matter of convention; what is important is that oppositely charged particles attract each other, whereas like-charged particles repel. Clearly, the strong force in the nucleus must be "stronger" than the electric force of repulsion between the protons, otherwise the atom would simply blow itself apart.

There are many important features that arise from considering the number of these particles in a given atom. Chemistry and biology in a broader context, are based on the observation that the number of protons in an atom determines its position in the periodic table, and therefore dictates how the atom behaves in the physical world. When we introduce the concept of mass, we could be talking about one of several different things, depending on the situation, but if you think about this for a moment, you'll notice that we haven't really done much beyond simple number counting. We could take as a working definition that the mass of a clump of matter is a measure of how much of this substance we have, which could simply be the *number* of sub-atomic particles (mostly protons and neutrons, since electrons are so light by comparison) assembled within it.

In physics, the definition of mass is somewhat more precise, and it is directly related to a body's *inertia*. If you've ever gone ice-skating with someone, you'll remember that when you push your partner (i.e., when you apply a "force"), his motion will change: he will accelerate, meaning that his velocity continues to change until you stop pushing. However, it does take effort on your part to do this, for his body is resisting your push. Sure, on ice the resistance is not very big, but it does take more than a simple tap to produce a noticeable change in his motion. Further, you'll probably remember that it takes more effort to cause the same acceleration if you exert a force on two people at the same time, or a friend that is bigger than the first. This resistance to an applied force is known as inertia, and the mass—or more precisely, the *inertial mass*—is the force required to produce a particular acceleration. (In symbols, we write $m = F/a$, where m is the mass, F is the force, and a is the acceleration. This expression says that the inertial mass is the force divided by the acceleration.) A commonly used unit of inertial mass is the *gram* or *kilogram*, but it could with equal validity be quantified in other (perhaps less practical) units, such as "number of humans required to produce an acceleration of 2 meters per second per second."

The resistance you feel when you push is not the frictional effect arising from the skate scraping the surface of the frozen pond. The inertia of a body persists even when a force is applied to it in vacuum, free of other impediments, and it's not difficult to see why the two definitions we introduced in the previous paragraph are similar. If we have twice as many sub-atomic particles in one clump as we have in a second, then the inertial mass of the former should be twice that of the latter. We could say that we have twice as many grams. It would take twice as much force to accelerate the bigger object by the same amount as the smaller one. So two of us pushing two friends on the frozen pond will produce roughly the same acceleration as one of us pushing just one of them at a time.

The second way in which we become aware of mass is through a body's *weight*. In the British Commonwealth, the U.S., and certain other parts of the world, we would say that an object weighs a certain number of pounds. Weight, however, is not the same as mass; it is the mass multiplied by the acceleration due to the Earth's gravitational pull. In other words, "weight" is a force, and its value changes depending on how strong the pull of gravity is where we make the measurement. An object's mass is the same no matter where it is, on the surface of the Earth, on Jupiter, or in space. It is an intrinsic property, which as we have seen, can be thought of as the number of sub-atomic particles it possesses. The Earth's gravitational pull, on the other hand, decreases as you move farther away from it, and so you can lose weight by changing your elevation, even though your mass remains the same. You can also lose weight by living on the moon or on Mars (which are smaller and lighter than the Earth, and hence produce a weaker gravitational pull on their surface), but again your mass is the same.

The story doesn't end there, however. It turns out that there's something quite peculiar about an object's weight, a feature that was evident even to Sir Isaac Newton (1642–1727). We will soon realize that we must accept without explanation an observation that is best expressed as a question: "Why should a body's weight depend on its *inertial* mass?" Remember that the inertial mass has something to do with its resistance to an applied force, yet here we are starting to talk about it as an important ingredient in determining the strength with which one object (the Earth) attracts another (a rain drop).

This distinction between the two ways in which mass manifests itself, first by the resistance it exhibits when force is applied to an object and second by how it brings about a gravitational attraction between two clumps of matter, was the seed that gave rise to a revolution in the theory of gravity. It ultimately led to the development of general relativity, the most successful description of gravity that we have to date. We will continue this discussion later in the chapter, but we must first try to understand what the origin of inertial mass is. Without doubt, a massless universe would be a strange place in which objects would present no resistance to impulses, and any contact whatsoever would send them careening chaotically at light speed into a general maelstrom of randomly moving matter. A consequence of massless matter would be that condensations such as stars and planets would not occur; wintry landscapes with powdery blankets of snow would be unknown; Kandinsky would have never graced the world of art; and Mozart would not have written a single note of music.

So why then do objects resist the influence of a force to move them? Without necessarily answering this question directly, Newton (see figure 3.1) settled on the idea that there must be an absolute space and time. Born in a manor house to a family of farmers in 1642, the same year that Galileo Galilei (1564–1642) died, Newton quickly displayed a rare depth and clarity of thought that led him, in 1687, to publish one of the most influential books ever written—the *Philosophiae Naturalis Principia Mathematica*, or *Principia* for short. In it, he analyzed the motion of a body acted upon by a force directed toward another object, and he applied the results to orbiting bodies (such as planets around the Sun), projectiles (such as stones catapulted from the surface of the Earth), pendulums, and the free-falling motion of apples. His international stature as a scientific leader has been sustained for hundreds of years, but like any revolutionary thinker, Newton had his detractors and fierce competitors. One of the more revealing anecdotes of his life involved the conflict with his main rival and nemesis, Robert Hooke, a leading member of the Royal Society, who criticized Newton's work, but at the same time complained that the *Principia* and his work on optics (another of Newton's monumental efforts) were plagiarized from him. Many believe that Newton's well-known statement, "If I have seen further than most it is because I have stood on the shoulders of giants," was his taunting response to Hooke, who was a very short man.[5]

Let us see what motivated Newton to propose the existence of absolute space by considering the following situation. Imagine that we're sitting in the passenger car of a train and that all the

---

[5]See for example Westfall (1981).

Figure 3.1 A painting of Sir Isaac Newton by Godfrey Kneller. (Reprinted with the kind permission of the Trustees of the Portsmouth Estates. Photographed by Jeremy Whitaker)

windows are shut so that we cannot see outside. Can we nonetheless tell if the train is moving at constant speed, or whether it is accelerating, or perhaps even decelerating? The answer, of course, is yes! If you hold a chain with your hand at one end and let the other end dangle freely, the chain will move to an oblique position backward when the train is speeding up, and it will move to an analogous position forward when the train is slowing down. It will hang directly downward when the train is moving at constant speed. Using Newton's definition, we would call the passenger car with the chain hanging directly downward an *inertial frame of reference*, the other cars being *noninertial*, or accelerated frames. What then determines which passenger cars are inertial and which aren't, or equivalently, how does the chain "know" when to exhibit resistance to the acceleration? Newton posited that there must exist absolute space and time, and that the inertial frames of reference (the passenger cars with the chain hanging straight down) are those that are either at rest in absolute space or in a state of uniform motion relative to it. In the *Principia*, Newton described the situation as follows:

> Absolute space, in its own nature and with regard to anything external, always remains similar and unmovable. Relative space is some movable dimension or measure of absolute space, which our senses determine by its position with respect to other bodies, and is commonly taken for absolute space.

Unfortunately, this simple idea does not work (for reasons that will become clearer when we introduce the special theory of relativity), and it gave rise to some spectacular debates in both scientific and philosophical circles.[6] In particular, the Austrian philosopher Ernst Mach (1836–1916) argued that the resistance, or inertia, exhibited by matter when a force is applied to it is "produced by its relative motion with respect to the mass of the Earth and the other celestial bodies." What Mach envisioned with this principle was that any object experiences an interaction with the rest of the mass in the universe that produces a resistance when it is pushed. It's as if every clump of matter is connected to every other clump by rubbery bands, which tug at it when the clump is pushed or pulled. Well, Mach's principle is not quite right either, although the current thinking on the nature of inertia does have

---

[6]See for example Alexander (1956).

some of its elements. As we will see, the viability of general relativity implies that there ought to exist frames of reference (like our passenger car) in which all the gravitational effects of the external universe vanish, so a particle's inertia cannot be due to matter *outside* of this frame.

Instead, the most fashionable idea currently being pursued by physicists is one that is suggested almost as an afterthought by the so-called Standard Model of nature. In this elegant description of the constituents of matter and their interactions with each other, the principal particles that make up the atoms and molecules in our bodies, the parchment of the Magna Carta and the glorious California sunsets over the Pacific, are all thought to be simple constructs employing a set of fundamental entities known as quarks and leptons (see table 3.1). The latter comes from the Greek word "leptos," meaning "small one"; the electron is a lepton, as are other light particles. The various combinations of the available flavors of these give rise to the multitude of sub-atomic species, such as the proton and neutron. To complete the picture, the Standard Model also incorporates "messengers" to carry the force between them—without these, no particle would be "aware" of anything else in the universe, and we would not be here to ponder these questions. Light is a field of many messenger particles known as photons, which transmit the force from one charged particle (say an electron) to another (such as a proton). The catch is that the Standard Model only works as long as there exists an underlying symmetry that makes all of these particles *massless*. So how does a massless theory help, given that matter does in fact have mass? Enter the Higgs particle.

The idea behind the Higgs mechanism is rather ingenious, and the accumulating evidence suggests that it may in fact be correct, at least in part. What's intriguing about it is that the manner in which mass is "acquired" by particles has a little in common with the concept of an ether from centuries long past, and maybe with Mach's principle as well. Starting with Newton, and continued by his followers into the latter part of the nineteenth century (see, for example, Maxwell 1954), physicists thought of light as either being made up of corpuscles (which we now call photons) or as being

Constituents of Matter		
First generation	**Second generation**	**Third generation**
u	c	t
d	s	b
ν_e	ν_μ	ν_τ
e	μ	τ

Table 3.1 The constituent particles in the Standard Model of Nature. The quarks, from which particles such as the neutron and proton may be built, are: **u** = up; **d** = down; **c** = charm; **s** = strange; **t** = top (or truth); and **b** = bottom (or beauty). The proton is a combination of two **u** and one **d** quarks, whereas the neutron is made from one **u** and two **d** quarks. The lower two rows comprise the family of leptons, which include the electron (**e**) and its neutrino (ν_e).

carried by waves undulating in an all-pervasive medium with very "ethereal" qualities. Searches for the ether produced null results and, at any rate, its existence is inconsistent with what we have come to learn from special relativity.

Nonetheless, although we no longer believe in the reality of an ether, there now exists something of a modern version. Suggested by the Scottish theorist Peter Higgs in the 1960s, the Higgs particles make up a virtually invisible field that pervades all of space—hence the loose analogy with the erstwhile ether. But it is not completely invisible because it does in fact interact with other particles. An electron must plow through this field when a force pushes it. The fact that the Higgs field clings to it (the term "cosmic molasses" has often been used in this context) means that the electron feels a drag, a back reaction that appears as a resistance to the applied force. We interpret the resulting inertia of the electron as a "sluggishness" that we quantify as mass.

Another way to see this is to imagine ourselves running late for a train at New York's Grand Central Station. If the station were empty of people, we could run unimpeded to our destination. Unfortunately for us, it's now peak hour, and the platforms are teeming with thousands of visiting Scotsmen, who for some reason are all named Higgs. In this situation, all of our effort is expended in an attempt to find a clear path to our train, but with the unavoidable frequent collisions between us and our fellow travelers, our progress is greatly delayed. The net effect is that the force we exert (through our feet moving on the pavement) is encountering a resistance from the frequent bumps with other people, which gives rise to an inertia. In the framework of the Standard Model, physicists often call this effect a breaking of the symmetry among the particles which would otherwise have zero mass. If this mechanism is correct, the Standard Model is viable because the Higgs field creates the massive world we know from a selection of zero-mass particles.

The Standard Model has produced several remarkable successes (and Nobel prizes) over the past few decades, with predictions that match experimental data with startling precision. This theory, however, is incomplete, for it does not include gravity (more on this later). In addition, like every other theory in science, it also does not "explain" anything; it merely describes what we see. The problem is that with the passage of time, the set of unknowns (such as the number of fundamental particles) seems to grow, suggesting that there may be a yet more fundamental description of nature still hidden below the surface. Once the Higgs is found, we may be able to see through the veils of ignorance and peer into a realm about which we have not yet even begun to contemplate. There are currently thousands of physicists and engineers around the world engaged in a systematic search for the elusive Higgs (see figure 3.2), and it is anticipated that a discovery will be made before the year 2010 at the Fermilab Tevatron in Illinois and the Large Hadron Collider (LHC) at the CERN laboratory in Geneva. In fact, we may not even have to wait that long, for at the time of this writing, CERN seems to be on the verge of uncovering the Higgs using its aging Large Electron Positron (LEP) collider.

3.2 DEVELOPMENT OF A GRAVITY THEORY

"Well, in my left hand, I have a feather; in my right hand, a hammer," says David Scott, "and I guess one of the reasons we got here today was because of a gentleman named Galileo ... who made a rather significant discovery about falling objects in gravity fields." This scene is being played out toward the end of the (northern) summer of 1971 (figure 3.3). Before the camera stands Commander Scott, who is about to conduct a dramatic demonstration next to the recently landed "Falcon," the lunar module from the Apollo 15 mission. He is about to carry out an Aristotelian version of Galileo's famous experiment with falling objects. Holding the feather (from a falcon, of course) and hammer between the thumb and forefinger of his left and right hands, respectively, Scott releases them simultaneously and watches as they fall side by side, hitting the ground at virtually the same time. "How about that!" rings in a voice from the background. "Which proves that Mr. Galileo was correct in his findings," effuses Scott. In the Apollo 15 Preliminary Science Report, this amusing incident is reported as follows: "Within the accuracy of the simultaneous release, the objects were observed to undergo the same acceleration and strike the lunar surface simultaneously, which was a result predicted by well-established theory, but a result nonetheless reassuring considering both the number of viewers that witnessed the experiment and the fact that the homeward journey was based critically on the validity of the particular theory being tested."

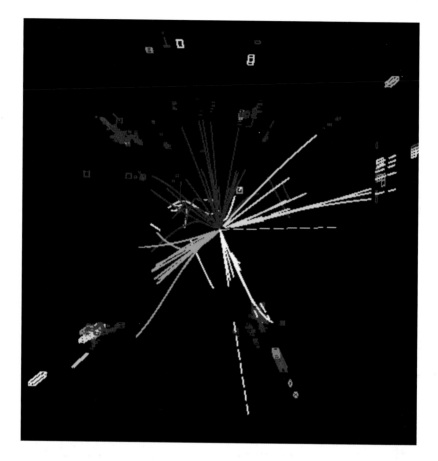

Figure 3.2 Jets of particles shoot out from an electron–positron impact that might have spawned a Higgs boson. Jets one and two (in red and white) could indicate the decay of a Higgs. DELPHI, which produced this result, is a Particle Physics experiment at the CERN laboratory in Geneva, Switzerland. It studies the products of electron–positron collisions at the LEP (Large Electron–Positron Collider) circular accelerator, which until recently was working at the highest energies in the world. (Event display produced with code developed by the DELPHI Collaboration at CERN, Geneva)

Galileo's experiment leading to the discovery that bodies fall at a rate independent of their mass was quite revolutionary, overturning tenets held "sacred" since Aristotle's (384–322 B.C.) pronouncements that heavier objects fall faster than light ones. In

Figure 3.3 The moon's lack of atmosphere provided the ideal conditions for astronaut Dave Scott's demonstration of Galileo Galilei's Renaissance-era discovery that a body's gravitational acceleration is independent of its inertial mass. Several hundred years earlier, Galileo had conducted arguably the most famous demonstration of this result by dropping heavy balls from the leaning tower of Pisa. (The Hammer and the Feather (1986); courtesy of the artist, Alan Bean, himself a former astronaut)

the Earth's atmosphere, the feather would indeed fall more slowly, but only because influences in addition to gravity are at play. In this circumstance, the air resistance depends on an object's size (the bigger it is, the more air it has to push out of the way), not its mass. Galileo avoided such complications by using objects for which these influences are negligible. By sliding weights down an inclined plane (not to mention dropping them from high places), he was able to eventually convince even the most ardent followers of Aristotle that a body's mass had nothing to do with the rate at which it fell. The biographer Stillman Drake (1981) notes that the Leaning Tower of Pisa story is probably true, only because Galileo (see figure 3.4) may have needed such a dramatic, theatrical demonstration to capture the public's attention. In this well-known tale, Galileo dropped a variety of lead balls from the tower, showing that they all hit the ground at the same instant, even though they clearly move faster and faster the further they drop, that is, even though they are being *accelerated* by Earth's pull.

This unexpected outcome was one of those historical turning points in our communion with nature that at once perplexes and inspires us. You may recall from the previous section that a body's mass is a measure of its inertia, of its resistance to an applied force. We thought about what happens to skaters on a frozen pond and decided that a heavy skater needs more force to make him accelerate at the same rate as a lighter one. Suppose we have two balls, one with twice the lead content of the other, and therefore twice as massive. When dropped from a tower in Pisa, or from the top of the Empire State Building, they nonetheless hit the pavement simultaneously, meaning that they accelerate by the same amount during their descent. But this requires that the *gravitational* force on the more massive ball be twice as big as the force on the lighter one to compensate for the mass difference. Our reasonable generalization from this leads to the surprising conclusion that the force on an object due to Earth's gravity must be proportional to its inertial mass—so 10 lead balls feel 10 times the gravitational force experienced by one alone.

Figure 3.4 Painting of Galileo Galilei by Domenico Robusti. (Reprinted with the kind permission of the National Maritime Museum, London)

Better still, we've all heard the saying that for every action there is an equal and opposite reaction (this is actually Newton's third law of motion), meaning in this case that if the Earth is pulling the lead ball with a certain force, the ball in turn is pulling the Earth in the opposite direction with equal potency. By analogy with the force felt by the ball, this means that the force of attraction must also be proportional to the Earth's inertial mass. In other words, two Earths together should feel twice the pull from the ball. The gravitational force, it seems, must be proportional to the inertial mass of *both* objects.

No one knows why the strength of gravity has anything to do with the inertial mass. Newton was mystified by this connection, as have been countless others since his time. Yet as we have already anticipated earlier in this chapter, a remarkable enlightenment follows if we accept it. But we digress. There is more to discuss on the classical thinking on this subject.

What we have considered so far is the behavior of objects on the surface of the Earth where the conditions are relatively uniform. Gravity's influence extends much farther than that—to the planets in the solar system; to the glistening spirals in galaxies; ... to the edge of the universe. But as the distance grows, the attraction diminishes. The Sun, far more massive than the Earth, ought to win any struggle with its planet for gravitational influence over our bodies, the conditions being otherwise equal. The fact that we remain anchored to the surface of the Earth rather than being drawn toward Apollo's fiery pit is due entirely to our proximity to the former and remoteness to the latter.

It was Newton who in 1665–1666 first deduced the inverse-square law based on the behavior of the moon and planets in the solar system. The moon, he argued, is 60 Earth radii away from the center of the Earth and it falls toward the latter a distance of 0.0045 feet per second. If the gravitational force obeys an inverse-square law, then an apple in Lincolnshire, which is correspondingly only one Earth radius away from the center, should fall 3,600 times 0.0045 feet, or about 16 feet in 1 second, which is about right.

He became fully convinced that this dependence on distance must be correct when 20 years later he proved that planets moving under the influence of an inverse-square law force would obey all the empirical laws deduced earlier by Johannes Kepler (1571–1630). With meticulous care and a life-long devotion, Kepler had by then established the harmony of the heavens by charting the motion of our planetary neighbors as they wandered around the Sun.

In the *Principia*, Newton encapsulated these ideas into the law of universal gravitation, about which he wrote:

... all matter attracts all other matter with a force proportional to the product of their masses and inversely proportional to the square of the distance between them.

This law explained many previously unrelated phenomena, including the eccentricity of cometary orbits (i.e., their noncircular trajectories), the tides and their variations, the precession of the Earth's axis, and the perturbation of the moon's motion by the Sun's gravity. It also created several philosophical conundrums, about which the continental scientists at the time had much to say. They could not accept the idea of action-at-a-distance, and sought alternative views in which forces act through contact. The idea of action-at-a-distance was not very palatable to Newton either. His law of gravity does not say how long it takes for the gravitational force between two bodies to establish itself, rather, it implies that the force is instantaneously present when the bodies are brought into the picture. Writing to Richard Bentley, he expressed his distaste for such a concept in this way:

It is inconceivable, that inanimate brute matter should, without the mediation of something else, which is not material, operate upon, and affect other matter without mutual contact; ... That gravity should be innate, inherent, and essential to matter, so that one body may act upon another, at a distance through vacuum, without the mediation of anything else, by and through their action and force may be conveyed from one to another, is to me so great an absurdity, that I believe no man who has in philosophical matters a competent faculty of thinking, can ever fall into it.

These objections notwithstanding, the Newtonian law of gravity withstood the test of time, at least until the latter part of the nineteenth century, when certain small, but irrefutable, inconsistencies started to unravel the smooth fabric of classical physics. Not only were scientists grappling with the deepening mysteries associated with how gravity works, but new evidence was beginning to suggest that the theoretical underpinnings themselves were only partially valid. Light, it turns out, behaves in a rather odd manner. Culminating with the Michelson-Morley experiment (1887), which measured how fast light travels along the direction of the Earth's orbital motion and transverse to it, physicists learned that the speed of light has the same value no matter who measures it.

There is something quite irregular about this finding, bequeathing a legacy of discomfort that threatens to dismantle our notions of distance and time. If two people moving relative to each other claim to infer the same velocity for light, the distances they measure in ascertaining how far it travels in a given time cannot be the same. Distance is a measure of how much space there is between two points, and a duration corresponds to how much time has elapsed between two events. Speed, being the ratio of distance over time, is therefore intimately connected with the properties of space and time. Therefore any empirical evidence that defies our intuitive concept of speed, such as the constancy of the speed of light, must surely be hinting at the fact that the measurement of distance, or time, or both, depends on who is making the measurement—odd indeed.

Take the case of a driver in a Ferrari traveling at 180 miles per hour on the autobahn, with serious intentions of overtaking a McLaren pacing along at a leisurely speed of only 170 miles per hour. The driver of the McLaren sees the Ferrari moving past her at $180 - 170$ (i.e., a mere 10) miles per hour. Quite reasonably, the Ferrari looks much faster from the side of the road (where we see it zipping past us at 180 miles per hour) than from the perspective of another vehicle moving in the same direction. However, if we were to send a pulse of light past the two drivers, not only would it overtake their cars, but they would conclude that the pulse is whizzing by them at exactly the same speed (approximately 670 million miles per hour), regardless of how fast they themselves are moving relative to us. Nature is telling us that the Newtonian worldview breaks down when very high velocities are involved, and since this is to be expected for matter falling into the maw of a very compact, massive object, Newton's law of gravity is simply not precise enough for us to conduct a study of the behemoth lurking at the center of our galaxy. However, before attempting to bring the theory of gravity into correspondence with the empirical knowledge gained in the twentieth century, it is essential for us to first understand how measurements of distance and time must be handled in order to correctly interpret a particle's acceleration in response to an applied force. Our task, daunting as it may seem, is to re-invent the equation $m = F/a$ in a form consistent with light's bizarre properties.

3.3 EVERYTHING IS RELATIVE

In 1905, a young physicist named Albert Einstein (1879–1955) published three remarkable papers, the first of which would eventually win him the Nobel prize in physics (awarded in 1921 in his absence while on a voyage to Japan). Dealing with a perplexing property of the electromagnetic field, this work answered once and for all the question of whether or not light was corpuscular. Einstein (see figure 3.5) examined the phenomenon discovered by Max Planck (1858–1947), according to which light seemed to be radiated in discrete quantities or bundles, and was subsequently absorbed as such. We now call these quanta of electromagnetic energy photons, and recognize them as being essential members of the Standard Model (see section 3.1 above).

Ironically, it is not for this paper that Einstein is best remembered, but rather for the second, which deals with yet another important property of light. (Incidentally, the third paper had nothing at all to do with light, but was instead an analysis concerning the burgeoning field of statistical mechanics.) The seminal work that brought Einstein widespread fame proposed what is now called the special theory of relativity, an attempt (successful, as it turns out) to reconcile the surprising results of the Michelson-Morley experiment with the classical view that the laws of physics should have the same form for any observer in any frame of reference. The motivation for this proposal is that if we cannot be sure what different people are measuring for distances and time,

Figure 3.5 Albert Einstein in the patent office (1905), at the time he published his first paper on the special theory of relativity. (The image was provided by the Institute Archives at CALTECH, courtesy of the Hebrew University in Jerusalem)

then it is meaningless to talk of physical laws, such as $m = F/a$, which are based entirely on the assessment of how these quantities change in response to a force.

Even so, we must not lose sight of the fact that Galilean relativity serves us quite well in our everyday experience, such as when we are driving our Ferrari or the McLaren F1. Galilean relativity, which states that one observer can understand what is happening in another observer's frame of reference by using universally accepted lengths and time, is clearly correct when the observers (and their frames) are moving relative to each other with speeds much smaller than that of light. In this realm, everyone agrees with the distances and intervals they measure, at least within the accuracy of their devices. So Galilean relativity should still be a sound basis on which to build a *new* relativity theory that is valid for all speeds.

Once again we are faced with one of those situations where nature is telling us something we don't understand, but the acceptance of which leads us to a greater enlightenment. To this day, no one knows why space and time behave the way they do (though some may claim otherwise), but all the indications are that, viewed from different frames of reference, the ratio of distance over time for light always has the same value, even though individually these quantities change. Go figure. In the words of I. I. Rabi, who was reacting to the discovery of an unexpected particle, "Who ordered that?"

Well, we've got to deal with this situation, as did Einstein. He was not the first to propose all the components of special relativity theory, but his enduring contribution was to unify important parts of classical mechanics and electrodynamics (essentially, the behavior of light). Special relativity begins with two postulates: (1) that the laws of physics are the same for all nonaccelerating observers, and (2) that the speed of light in vacuum is independent of the motion of all observers and sources, and is observed to have the same value. We will accept the fact that the second postulate forces us into a situation for which distances and time are different for different observers. The first postulate implies something quite profound about space—it states that we are not able to infer an absolute velocity relative to it. Imagine we are dining on an airplane traveling at 200 miles per hour relative to the ground. If it were to accelerate to a different velocity, of say 400 miles per hour, we could resume our meal without being able to distinguish this velocity from the former.

The second postulate is even more radical than we have let on thus far, since it requires not only a lack of uniformity in the determination of physical quantities, but more importantly, it demands an *intertwining* of the measured time and space. In the process of unraveling this mystery, we will have to ask ourselves whether we really understand what time is. If the velocity of light is independent of the relative velocity between observers, then the possibility of a universal concept of simultaneity must be ruled out. When one observer says that two events occur simultaneously, while a second sees that one precedes the other, clearly the inferred time "flows" differently for the former compared to the latter, but in a way connected with the behavior of space. Again, suppose we are sitting on the airplane, this time right at the midpoint between its nose and tail. If we were to turn on a beacon of light at our position, we would see the light pulse go out in both directions, front and rear, reach the end points where we have cleverly placed mirrors, and then see the light reflect off these mirrors and reach us at exactly the same time. This is what we mean by simultaneous events: that the times at which the light pulse arrives at the end points, as measured by us, are identical.

Our friend standing on the ground sees something quite different. If the plane passes overhead, he sees that the light pulse traveling forward takes longer to reach the nose of the plane than the other pulse does to reach the rear. The delay occurs because the plane is moving forward, so by the time the light pulse reaches the mirror, the nose has moved beyond where it was when the beacon was turned on. Of course, the reverse is true of the tail mirror, for the rear of the plane has also moved forward (toward the point where the light pulse originated) while the light pulse is traveling backward. The net result is that our friend on the ground sees the light pulse reach the rear mirror *before* it reaches the front. These events (the two reflections of light) are therefore not simultaneous according to him. There is no paradox in this diversity of outcomes. We are simply witnessing the effect of a constant speed of light, as expressed in the second postulate. If all observers agree that the speed of light is invariant, then when we measure intervals of time, and by association, distances traveled by light, observers moving at different velocities relative to each other must see events occurring at different places and times.

Many of us have difficulty grasping what these concepts really mean because (truth be told) we are really Newtonian at heart. That shouldn't be surprising since the world in which we move is described adequately by Newtonian mechanics, and we develop an intuitive understanding within this framework that serves us (and evolution) well. What we mean by this statement is that we don't think deeply about the nature of time, but rather rely on our visceral sense that time is like a fluid that "flows" at a steady rate, uniformly and in an absolute manner for everyone and all things. Beyond that, only a handful of philosophers have attempted to quantify its meaning, falling short due to imprecise guidance from the physical world. In the latter part of the fourth century, St. Augustine (354-430) wrote: "What, then is time? If no one asks me, I know what time is. If I wish to explain it to him who asks me, I do not know."

If we listen carefully to the language of special relativity, however, we do pick up certain helpful clues. Among them is the fact that when observers determine distances and intervals, they must make the appropriate measurements using devices, which might be something as simple as a ruler, or somewhat more complicated, like a mechanical clock. When we ascertain that a number of seconds have elapsed, this result is based on how many ticks on the clock we have counted. And there we go again! Just as we were reduced to an exercise in counting to determine the mass of a body (for example how many humans does it take to cause an ice skater to accelerate at 2 meters per second per second?), so too we must rely on counting something to uncover how much time has passed. We are therefore Newtonian in our thinking because we take it for granted (in the absence of any evidence to the contrary) that what we are counting is the same for everybody, that the clock produces identical ticks for every observer, anywhere in space.

The Michelson-Morley experiment, and many others that followed, show us that this is not the case, and that we must instead think of time as an objective expression of the changes that occur in the state of matter. Time exists in a certain region of space only because the constituents at that location are changing. So we infer that time has passed because the inner workings of the clock have caused its hand to move across its face. Time and movement are inseparable concepts. If all physical processes within a given region of space were to slow down to half their normal speed, time would correspondingly "flow" at half the rate with which it flows elsewhere. As a measure of how quickly things change, time would therefore stop if everything were to become frozen in place. One wonders if this is what the founders of Buddhism had in mind when they wrote that the aspirations of the devout followers should be to reach Nirvana, a state in which time ceases to exist. Certainly Aristotle seemed to argue for a tight coupling between time and change when he wrote: "Movement, then, is also continuous in the sense in which time is, for time is either the same thing as motion or an attribute of it."

For these reasons, and the viability of special relativity, we must adopt a view unlike that of Newton, who envisaged time as flowing unobstructed with a steady progression everywhere, whether or not matter is present. In his world, time is conceived as having an existence separate and apart from the physical universe; indeed, it would exist even if the universe did not. Instead, we should think of time as a measure of how many still images have passed when we view a motion picture. If each frame is a unit of time, then we can evaluate how much time has elapsed for the action in the film by counting the frames. If we slow the film down, the images (and therefore time) pass by us more slowly. In this context, we see that time for the world depicted in the film is intimately connected with the action taking place therein, and it ceases to exist altogether when the film is stopped.

We are compelled, therefore, to think of space and time in a unified manner, losing any semblance of division by advancing the terminology to the next level, in which we speak of the *spacetime* rather than the space and time. What affects space bears on the changes within it and hence also affects time.

In a moment of contemplation, we may well ask what any of this has to do with gravity, black holes, event horizons, and the galactic center. Let us try to collect our thoughts and see where we need to go from here. Gravity is associated with mass, or more precisely, inertial mass, which is a property of matter that can only be determined by sampling its response to an applied force. We can call the force whatever we want (one horse, two humans, or more commonly, a newton, in honor of the great thinker), but its effect—the acceleration—must be calculated using distances and times (for example 2 meters per second per second). Evidently, for us to use the concept of gravity, we must be able to say with confidence who is making these measurements, and how two different observers view the common world in which they exist. (After all we've been through in this chapter, let us not now veer off into an argument with Descartes over whether or not we truly do exist!) In the next section, we will characterize the effects of gravity even more directly in terms of the acceleration; by then, our involvement with special relativity will be inescapable. To invoke gravity we must rely on the structure of spacetime, so to answer the question we posed above, we need to know what effect special relativity has on distances and intervals of time.

Two of the most important consequences of special relativity are that lengths viewed from a moving frame shrink and times are dilated. As we said earlier, no one knows why space has this property, but if light is to have an invariant speed, this condition is a necessity. Whenever an object with mass is in motion, its measured extent in the direction of its motion looks shorter to us than it does to someone moving along with it. Moreover, the object shrinks by a greater amount the faster it moves and it would shrink to nothing if it were to reach the highest possible speed— that of light. However, only a person who is in a different frame of reference from the object would actually see this shrinkage; as far as the object itself is concerned, its size remains the same as measured in its own frame. This phenomenon is known as *length contraction*.

The phenomenon known as *time dilation* arises in a similar fashion, having to do with how the rate of change in a given frame is perceived by an observer moving relative to it. This effect manifests itself as a stretching (or dilation) of the time interval so that each tick of the clock in a frame moving relative to us takes longer than one tick in our frame. Revisiting the analogy of the motion picture, we would see a slowing down of the film displayed on a projector in motion. The still frames would flash by us ever more slowly as the projector's speed increases, stopping altogether when it moves about as fast as light. The dilation of time has been verified experimentally in a variety of ways. For example, neutrons are particles that decay rather quickly when they're isolated outside of the nucleus. They have a half-life of only 10.3 minutes, meaning that if you start with 100 neutrons, half of them will have disintegrated during this brief time. When they're produced with very high energy in particle accelerators, however, they travel very close to the speed of light, and as viewed by us in the laboratory, their lifespan is measured to be much longer than 10 minutes. In fact, the main subject of this book—the galactic center—is itself a source of very energetic particles, including neutrons, which transit across the 28,000 light-year distance to us. These particles are moving so close to the speed of light that a 10-minute lifespan in their own frame is for us dilated into 28,000 years!

It should not surprise us, therefore, to discover that a particle's inertial mass also changes as its velocity increases. Remember that the inertia is inferred from the acceleration due to an applied force ($m = F/a$), so if distances contract and times dilate, the perceived acceleration, which is a measure of how much an object's velocity changes in a given time, must itself diminish even under the influence of a constant force F. The ratio F/a, that is, the inertial mass, increases indefinitely as the speed of light is approached since a gets progressively smaller.[7]

---

[7]Let me here inject a word of caution for the more serious reader who will want to pursue this discussion in the technical literature. There are several ways of defining a particle's mass, though of course always describing the same physics. Sometimes, a particle's mass is taken to be the (fixed) value measured in its own rest frame, and not unreasonably, is then called the *rest* mass, m_0. The rest mass never changes, no matter how fast the particle is moving, since all observers agree that it represents the inertial mass in that one special frame. With this usage, the changing inertia is then characterized by a multiplicative function, γ, called the Lorentz factor, and everything we've said so far about

One of the most famous equations in history, $E = m c^2$, is a direct consequence of this phenomenon. In words, it says that the total energy acquired by a particle is proportional to its inertial mass. The presence of the constant factor c^2 (which is the speed of light squared) is there only to make the dimensionality on the left-hand side of this equation match that of the right-hand side; this factor never changes. The inertial mass, on the other hand, does change, and with it so too does the particle's energy. This equation comes about because the longer we apply a force to an object, the more its inertia and total energy increase, giving rise to the strict proportionality of energy to mass, written symbolically as $E \propto m$. The minimum occurs when the particle is not moving, since that corresponds to its smallest inertia. The energy would reach infinity should the particle attain light speed, which is clearly impossible since not even the entire universe has infinite energy. So objects with mass can never reach the speed limit, and light (as far as we can tell) must be massless in order to avoid this problem. Is there any experimental verification of this equation? One of the most dramatic demonstrations of the proportionality between inertial mass and energy has been the power and consequent destructive capability unleashed during a nuclear explosion. Because c^2 is such a large factor, even a tiny inertial mass represents a huge storage of energy, whose release has been witnessed many times prior to the test ban treaties negotiated by the nuclear superpowers.

3.4   THE PRINCIPLE OF EQUIVALENCE

Having lived through the inception of the space age, most of us are familiar with scenes of astronauts floating freely about their orbiting spacestation, bounding effortlessly from one wall to the next in seemingly random directions. One of them may release a pad, which remains stationary where she left it, while another squeezes a tube of juice, whose droplets form a straight-line tra-

---

m then applies to γ, since $m = \gamma m_0$. For a comprehensive treatment of relativity, see Anderson (1967), Weinberg (1972), and Wald (1984).

jectory toward their final destination—the open mouth of a fellow traveler. The astronauts sleep in all sorts of positions, feeling no stress on their bodies since their cabin is a world with *zero gravity*. Yet we will not see the spacestation break free from Earth's influence. Orbiting spacecraft simply do not float away, but the world inside them is gravity-free. What is happening?

In 1907, two years after the advent of special relativity, Einstein was writing a review on the new physics, which he called "invariance theory," when he began to wonder how Newtonian gravitation could be modified to make it consistent with the effects we described in the previous section. The idea that took form in his mind, described later by him as the "happiest thought of my life," was that an observer who is falling from the roof of a house experiences no gravitational field. Yet the observer, like the orbiting spacestation, is clearly not breaking free of Earth's influence.

We can all accept the fact that the observer is indeed falling. Although many of us may not recognize that the spacestation and all its contents are also falling toward the Earth, they certainly are. It so happens that they're moving sideways with just the right speed (about 17,400 miles per hour) to offset their motion downward. As the spacestation falls it moves laterally with sufficient speed to merely keep up with the Earth's curvature, and it therefore follows a circular orbit. A speed above 17,400 miles per hour would have it move sideways too fast and its trajectory would push it to a higher elevation; obviously a lower speed would allow the spacestation to fall to a lower level.

These "objects," the unfortunate homeowner, the spacestation, and the astronauts and their belongings, are all in *free-fall* within Earth's gravity. But thanks to Galileo's profound insight, we now understand that everything inside the spacecraft is falling at exactly the same rate, since the acceleration they experience due to Earth's pull is completely independent of their inertial mass. Be it a pen, a paper clip, or the flight commander, they all exist within a region where no one can tell if anything else is being pulled or if something is pulling them, since every object is accelerated in tandem with everything else. Astronauts in training can certainly appreciate the significance of this phenomenon during their sessions inside a modified KC-135 aircraft, in which they can experience brief periods of microgravity. Renamed the "vomit comet" by its hapless passengers, this vehicle creates 30 to 40 seconds of simulated weightlessness by flying along a parabolic trajectory, like the path followed by a baseball once it leaves the rightfielder's hand, or a cricket ball hurled toward the stumps from long leg. During this period, the jet plane and all its contents are falling freely, and so its interior behaves as if there were no gravity.

The profound implication of this phenomenon is that regardless of what actually causes the pull of gravity, its effect is entirely equivalent to a uniform acceleration throughout a given volume of space. Einstein explained it in this way: imagine that you are standing inside an elevator out in space, very far from any object that can generate a gravitational field. Now you and your belongings really are in a microgravity environment, but suppose that a rocket engine attached to the bottom of the elevator is turned on and it accelerates your capsule. Our everyday experience driving cars and flying in airplanes guides us correctly in this instance, and we expect our feet to feel a force "upward" from the floor of the elevator being accelerated in that direction. But inside the elevator, we can't tell if it's our capsule that is being accelerated upward, or if we and our belongings are being pulled downward. The two would be entirely equivalent, since in both cases we infer that everything inside the elevator is accelerating downward relative to it. Try releasing a pen now, and you would see it "fall" toward the floor, since you and everything else are being boosted in the other direction. Einstein called this the "Principle of Equivalence," about which he wrote:

> ... we shall therefore assume the complete physical equivalence of a gravitational field and the corresponding acceleration of the reference frame. This assumption extends the principle of relativity to the case of uniformly accelerated motion of the reference frame.

Einstein reasoned that since all gravitational fields vanish inside a free-falling frame, special relativity ought to apply to all measurements of distances and times within that frame. And in a leap of faith (or inspiration, or both), he argued that the two postulates of the special theory should apply even in cases where we compare the measurements of an observer in this frame with those of an observer far, far away, where the effects of the gravitational field are negligible. Thus, in an elegant and all-encompassing way, the theory of gravity has been merged with the framework of special relativity. Gravity is described by its equivalence to an accelerated frame, and its effects are thereby fully incorporated into the laws of physics via the properties of special relativity. This is the essence of the *general* theory of relativity.

Now that a way has been found to merge the physical consequence of a gravitational field with spacetime, it remains to develop a scheme for actually determining the gravitational field produced by the presence of an inertial mass. These are two complementary aspects of the overall picture. On the one hand, the Principle of Equivalence allows us to evaluate the impact of an existing gravitational field on its surroundings. To complete the job, we also need to know how to calculate the field produced by a known inertial mass distribution. For Newton, the analogous completion of the theory had been achieved when he realized that the force of gravity was proportional to the inertial masses of the two interacting bodies, and inversely proportional to the square of the distance between them. This theory is certainly adequate for most situations in which two clumps of matter exert a gravitational influence on each other, for example as is the case between the Sun and Earth. But this can't be completely correct under all circumstances. For one thing, there's the equation $E = m c^2$, which now starts to take on an additional meaning. Since energy and inertial mass are equivalent, does that mean that the force of gravity should also depend on other forms of energy within an object as well as its mass? This generalization would mean that simply using a body's measured inertial mass is not enough to determine its gravitational pull.

There's also that annoying business about "action at a distance," which is in direct violation of special relativity. It makes no sense for us to merge a Newtonian gravitational force with the framework of relativity if we don't at the same time restrict how the effects of that force are mediated through spacetime. In Newton's theory, the Earth experiences the Sun's gravity instantaneously, so that if the Sun were to suddenly deviate from its current position, the Earth would feel that motion concurrently. In special relativity, however, no influence can travel faster than the speed of light, so Einstein would argue instead that the Earth would feel that sudden motion only 500 seconds later, this interval being the length of time required for light to reach us from the Sun.

The idea that the effect of gravity itself should travel at the speed of light gained considerable experimental verification in the latter part of the twentieth century. In concrete terms, this means that contrary to the classical view that gravity should be felt instantaneously everywhere in space, its influence must instead be carried from point to point by a new type of radiation field—let's call it gravitational radiation. A direct measurement of the speed of gravity is not yet possible in the laboratory due to the very weak nature of the force, but indirect means based on the behavior of astrophysical objects are feasible, and some have produced satisfactory results. For example, the binary pulsar designated PSR 1913+16 has an orbit that is decaying, and this behavior is attributed to the loss of energy due to escaping gravitational radiation. The rate at which this energy is lost depends on the finite speed of propagation, and the orbital changes can therefore equivalently be viewed as a measure of this velocity. The fact that this damping occurs at all is a strong indication that something (i.e., the gravitational radiation) is leaving the system and that it must be doing so with a finite speed—otherwise, the whole collapse would occur instantaneously. The actual measurement confirms that the speed of gravity is equal to the speed of light to within an accuracy of 1 percent.[8]

---

[8]For a technical reference on this subject, see Damour (1987).

The span of eight years between 1907 and 1915 was an intellectually turbulent time for many physicists struggling to develop the methodology for calculating the attraction due to gravity. Several major figures of science had to pool their intellectual resources to produce the final, correct form of the equations that describe the gravitational field produced by matter. Taking into account the source of gravity due to energy as well as mass, and incorporating the properties of mediation of the field through spacetime, these equations describe how the force changes as one passes through a region containing matter. The change is more severe the greater the inertial mass content, less severe (or even zero) when passing through a void. Ironically, the first paper to present the results of these monumental efforts was not even that of Einstein himself. Five days before Einstein submitted his 25 November 1915 paper, *The Field Equations of Gravitation,* which gave the correct equations for general relativity, the great mathematician David Hilbert (1862–1943) had himself submitted a paper entitled *The Foundations of Physics,* which not only contained the correct field equations, but also some important additional contributions to relativity not found in Einstein's work. The two had shared ideas openly leading up to this point, and clearly each had influenced the other's thinking.

In a matter of months, Karl Schwarzschild (1873–1916) had succeeded in showing how these equations could be used to understand the gravity produced by a single compact object, such as the Sun or the Earth. A professor at Potsdam, Schwarzschild had volunteered for military service in 1914, and had been stationed on the Russian front when he received copies of Einstein's papers. He unfortunately contracted an illness soon thereafter, and died upon his return home. At the time, his work was considered to be purely theoretical, but subsequent studies with very compact objects (such as neutron stars, which carry as much mass as the Sun within a city-sized region) relied heavily on the gravitational field described by his equations. Schwarzschild's name is now rightfully associated with several distinguishing features of black holes, about which we will have much more to say in the following sections.

3.5   THE KEY PREDICTIONS OF GENERAL RELATIVITY

It is safe to say that the most peculiar aspects of general relativity trace their roots to the nature of light and the inferences we draw from its propagation in spacetime. Thinking about the issues we discussed above, we realize that if it weren't for the second postulate of special relativity (i.e., the fact that the speed of light in vacuum is independent of who is making the measurement), the equations describing the gravitational attraction produced by one clump of matter on another would be a restatement of Newton's simple (and empirically derived) inverse-square law. The two effects introduced by special relativity—the inclusion of energy along with mass as the source of gravity, and the finite propagation speed of the gravitational force—alter this law in very significant ways.

These modifications are certainly pertinent when we discuss the gravitational attraction between bodies that are emitting gravitational radiation, such as the neutron star and its unseen companion in the binary pulsar PSR 1913+16 to which we have alluded above. These effects can also manifest themselves at much smaller distances when the energy content in matter is large compared to its inertia, especially for objects that have collapsed to extremely high densities. For example, this neutron star has a mass of about one and a half Suns compressed within a region no bigger than the city of Chicago. Think about what happens when we use a pump to inflate a tire. The more we squeeze the air by compression, the hotter the pump becomes. Now let us imagine squeezing a Sun into a city-sized volume and we can begin to appreciate how hot these stars must be. Heat, like other forms of energy, contributes to the gravitational attraction.

Gravity's Effect on Light

But the two effects on light introduced by special relativity are only the beginning of what awaits to be unveiled, for the consequences of light's behavior in the presence of a gravitational field challenge even the most imaginative physicists. On 8 November 1919, the *London Times* proclaimed "The Revolution In Science/Einstein Versus Newton" in one of its headlines. Two days later, the *New York Times* echoed with "Lights All Askew In The Heavens/Men of Science More Or Less Agog Over Results Of Eclipse Observations/Einstein Theory Triumphs." The commotion around the world was being fueled by the solar eclipse that year, during which an experiment confirmed that light rays from distant stars are deflected by the Sun's gravity in the amount predicted by general relativity.

In truth, general relativity does not "predict" that light paths should be bent by gravity, but the Principle of Equivalence allows for this possibility, whose confirmation by the solar eclipse experiment in 1919 was as much of a surprise as it was a turning point in our interpretation of how gravity works. The Principle of Equivalence correctly states that the effects of gravity are equivalent to a global acceleration of the underlying frame of reference. Standing inside an elevator, we can't tell whether we and everything around us are being "pulled" down by Earth's gravity, or whether the elevator is being accelerated upward. The discovery by Galileo and Newton that the gravitational force on an object is proportional to its inertial mass made this possible since it "explains" why the acceleration is the same for all matter.

This reasoning, however, has very little to say about particles with *zero* mass—such as the photon constituents of light. Perhaps the behavior of light could after all be the distinguishing factor between the effects of gravity and those of an accelerated frame. In other words, suppose we grant that for some unknown reason the gravitational force on an object is proportional to its inertial mass. Then the equivalence between the effects of gravity and those of an accelerated frame could apply to all *massive* objects but not necessarily to the *massless* ones. In that case, a light path would appear bent in an accelerated frame, but not in a gravitational field, since a ray of light propagating horizontally across the elevator (as seen from outside) would appear to be curving downward to the upward-moving observer inside.

The excitement generated in November of 1919, which turned Einstein into an overnight celebrity, was ignited by the realization that everything—not just massive particles—is affected by gravity. Massless or not, light follows a curved trajectory, just as the baseball does when it is thrown toward home plate from left field. The presence of a massive object, it seems, changes the spacetime around it to make everything follow a curved path. Be they cricket balls, protons or photons, they all accelerate toward the source of gravity at the same rate, inspiring in muselike fashion the thought that spacetime itself is distorted to make this happen. After all, if the gravitational acceleration is identical for all things, then it ought to be due to a modification of the spacetime through which everything is moving.

This idea (now experimentally verified; see figure 3.6)—that a gravitational field ought to bend the path of light—sits at the heart of everything we have to say in this book. It is so intriguing that its genesis actually preceded general relativity by over a century, though forged in the context of Newtonian mechanics. In 1783, the Reverend John Michell (1724–1793) adopted the prevailing corpuscular theory of light and imagined its bundled constituents trying to escape the Newtonian gravity of stars. Michell was born three years before the death of Isaac Newton, and became a well-known British geologist and astronomer, later regarded as the "father of seismology" in his study of earthquakes. He is also credited with the idea of binary stars, and the demonstration of an inverse-square law in magnetism. Rocks thrown upward, he reasoned, reach a maximum height dictated by their initial velocity, and then they fall back down to Earth. Newtonian mechanics shows us that the faster the rocks begin their upward motion, the higher they go. So if we project them with sufficient speed from the ground, they can even break free of Earth's gravity and escape our planet's clutches. Not unreasonably, the minimum velocity required for a rock (or a space vehicle) to leave the Earth is referred to as the *escape velocity*, whose value increases in proportion to the inertial mass of the object (in this case the Earth) producing the gravitational attraction.

Figure 3.6 The galaxy cluster Abell 2218 provides us with a modern version of the light-bending excitement generated by the 1919 expedition to the west coast of Africa. Abell 2218 lies approximately 2 billion light-years behind the constellation Draco. Deep Hubble images reveal many beautiful arcs surrounding its center. Most of these arcs are actually gravitationally focused images of a single galaxy, about 5–10 times more distant (and behind) the cluster. The cluster is so massive that its enormous gravitational field deflects light rays passing through it, much as an optical lens bends light to form an image. This phenomenon, called gravitational lensing, magnifies, brightens, and distorts images from faraway objects. In this particular case (which is rare) the light is bent so strongly that the image is sheared out into an arc. In addition, the strength of the lens is such that light passing to either side of the cluster gets bent toward the observer and multiple images form. (Courtesy of A. Fruchter at the Space Telescope Science Institute, and NASA)

Michell argued in a paper published by the Philosophical Transactions of the Royal Society that if a star was sufficiently massive, its escape velocity would have a magnitude exceeding even the speed of light, which being composed of particles, would then slow down and fall back to the surface. These stars would therefore be unobservable and he coined the term "dark star" to vividly portray this peculiar property. (Incidentally, this discussion appears to have been the first mention of the possible existence of dark matter in the universe.) Not aware of Michell's work, the great French mathematician, astronomer and physicist Pierre Laplace

(1749–1827) proposed similar ideas to those of Michell in his famous 1796 paper "Exposition du Système du Monde." In it, he showed that some stars could have a gravitational field so strong that light could not escape, but would instead be dragged back onto the star.[9] Of course, the experiment of Michelson and Morley in 1887 proved that light always travels at a speed of 186,000 miles per second, no matter where it originates. So this simple mechanistic view of what happens to light as it "climbs" out of a gravitational well is not an accurate depiction of how nature actually operates.

Imagine, then, the sobering impact of the solar eclipse expedition to the west coast of Africa over 100 years later, when the world was stunned with the spectacular manifestation of gravitationally induced light bending. It was now established without any lingering doubt that gravity does indeed affect the path of light, just as it does everything else in the universe and, moreover, it was apparent that the degree to which a ray of light deviates from a straight-line path is in proportion to the gravitating object's mass. What happens, the inquisitive mind would ask, when the mass of a star is indeed so large that the induced curvature on the trajectory of light prevents it from escaping? According to special relativity, founded on the principle that nothing travels faster than light, everything within the star must then be forever entombed.

By the 1930s, J. Robert Oppenheimer (1904–1967) and his collaborators had begun to investigate in earnest the evolution of stars beyond the point where much of their nuclear fuel is spent. While on the main sequence, meaning the period during which the star burns hydrogen-rich fuel at a steady rate, the ashes collect within its core, which forms a growing reservoir of heavy elements such as iron, carbon, and oxygen. The star can retain this balance for billions of years because the energy released from nuclear burning can sustain buoyancy in the hot gas against the inward pull of

---

[9] A detailed account of Laplace's contributions to mathematics and astronomy may be found in Gillispie (1997). Additional material on the history and physics of "dark stars" appears in Thorne (1995).

gravity. But the fuel eventually runs out. Oppenheimer, best re-membered publicly for his contribution to the wartime Manhattan Project that developed the atomic bomb, is also admired scientif-ically for having been the first to show that the end point of this stellar evolution can result in the catastrophic collapse to some-thing approaching Michell's and Laplace's dark star, though with significantly more profound implications. Thus, 24 years after the foundation of general relativity, Oppenheimer was challenging physicists everywhere to accept the apparently absurd notion of dead stars collapsing without limit to an indefinitely small size and infinitely large density, a point called a "singularity." His cal-culations, incorporating the effects of general relativity, revealed that evolved stars with ashen cores more massive than three Suns, have no means of preventing the inexorable inward pull of gravity. Research groups working in the east and the west had different names for these singularities: the Russian scientists called them *frozen stars*, while western physicists were calling them *collapsed stars*. Neither designation fully described these objects, however, and it came (in 1967) to John Archibald Wheeler to coin the term *black hole*, which was quickly and enthusiastically adopted by the astrophysical community. (Only the French found exception to the term "trou noir," which nonetheless found mainstream accep-tance in due course.)

Since then, computational refinements in the life and death saga of stars have produced a greater variety of possible outcomes. We now know that some stars burn their nuclear fuel so rapidly—for example, the main sequence phase of a 40-solar-mass star is over in several million years, compared to that of our Sun which won't reach old age until several *billion* years hence—that they enter a protracted existence engulfed in their own embers (see fig-ure 3.7). What remains of these objects living in the fast lane is a very compact core, whose constitution depends on whether its mass is greater than or less than an important limit known as the Chandrasekhar mass, roughly 1.4 Suns. Above this limit, gravity squeezes most of the electrons and protons together, leav-ing a residue of neutrons, hence the designation of these spent

Figure 3.7 The Hubble Space Telescope captured this beautiful image of large, billowing gas and dust clouds expanding away from the supermassive star Eta Carinae. At this site, a giant outburst occurred about 150 years ago, becoming one of the brightest stars in the southern sky. Releasing as much visible light as a supernova, this star nonetheless survived the outburst. It is estimated to be about 100 times more massive than the Sun, and is expected to produce a full-blown supernova sometime in the near future. That explosion will eject most of the outlying gas in Eta Carinae's tenuous envelope, but since this star is so massive compared to many others that reach this catastrophic end, the ashes in its core may very well produce a small black hole when the disruption finally occurs. (Courtesy of NASA, Jon Morse at the University of Colorado, and Kris Davidson at the University of Minnesota)

objects as "neutron stars." Stars that are smaller than about eight Suns produce ashen cores below the Chandrasekhar limit, and these are known as "white dwarfs." Supported entirely by the pressure of its trapped electrons, a white dwarf contracts gravitationally to about the size of the Earth, and survives in this form thereafter, cooling all the while. Larger stars form more massive cores of burnt gas and when their supply of fuel is fully consumed, they collapse rapidly with an explosive ejection of the outer layers, which we call a supernova. In this case, the remnant is a neutron star, in which the supporting pressure is provided by the neutrons.

Neutron stars are not the most massive objects in the universe, but their density and gravitational field intensity are second only to those of black holes. Given that within them a little over one Sun's worth of matter is compressed into a city-sized volume, the density of these stars is so extreme that a teaspoon of neutron-star material is equivalent to the world's entire population, weighing over 100 million tons on the Earth's surface. The gravitational pull of these objects is so strong that a marshmallow dropped onto the neutron-star surface would hit it with sufficient power to release as much energy as a typical nuclear explosion. But even the dense concentration of neutrons is insufficient to stop the force of gravity when the mass of the core exceeds about three Suns, as Oppenheimer had surmised over 60 years ago. For these moribund objects, the inward pull is irresistible and the collapse continues rapidly toward an apparent singularity of infinite density.[10]

Gravity's Effect on Time

Communities around the world enforce a speed limit on their roads. Attempts to limit the acceleration, however, are rare. Thus, while the driver of the Ferrari may find it frustrating that he cannot often test the postulates of special relativity with sheer speed, he can nonetheless savor the tangible exhilaration of raw power and the *rate* of increase in speed thereby induced. The acceleration is independent of velocity; whether the Ferrari is starting from rest, or burning tar on the autobahn, an acceleration will

---

[10]Read more on the history of the black hole idea in Wheeler (1999).

have the same impact. Accelerating at 30 miles per hour per second, the car will have a speed of 60 miles per hour after 2 seconds if it starts from rest. Its speed would be 260 miles per hour if instead it was already traveling at 200 miles per hour before the acceleration. In either case, the change in velocity over a given time interval is the same.

Viewed from a distant vantage point, a spaceship falling in the gravitational field of a *black hole* accelerates at a rate commensurate with its height above the singularity, regardless of how quickly it is moving. (Inside, of course, everything is falling freely and the travelers have no sense that a gravitational field is pulling them in.) In special relativity, the distortions to intervals of time and distance are entirely dependent upon the relative velocity between two different observers: the astronaut inside the ship and someone far away. General relativity, however, introduces additional distortions that constitute more than a mere subtlety. To be sure, the special relativistic length contraction and time dilation must still apply to any pair of reference frames, whether or not they are imbued with a gravitational field, but the acceleration does something new—it causes clocks to run more slowly in the presence of gravity. The reason for this effect is wedged firmly between the meaning of time and its dependence on change.

Time passes when something is changing. To measure a second on a clock, the hand must turn one-sixtieth of a complete cycle across the clock's face. Let us suppose that the clock is inside an elevator accelerating upward at a rate of 2 feet per second every second. Regardless of how fast the elevator is moving at this instant, during the next second on the clock, its speed will have increased by 2 feet per second. As long as the elevator is accelerating, there is no way to avoid the fact that its velocity relative to the outside world is different at the end of that elapsed second compared to its value at the beginning. Undaunted, we instead let the clock run for half a second. This time, the elevator speeds up by 1 foot per second, but again the velocities are different at the beginning and at the end. Unfortunately, no matter how much we shrink the interval of time, the starting and ending velocities are always different—that's the nature of acceleration.

The same situation applies to our travelers inside the spaceship approaching the black hole (or any source of gravity, for that matter). They look at their clock and measure an interval of time. But in the act of allowing time to pass, they have crossed into a frame of reference moving even faster relative to us than they were before. Thus, according to special relativity, there should be an additional time dilation associated with this increase in speed. The effect is greatly magnified when the acceleration is so great that even a minute interval can bring the magnitude of the ship's velocity uncomfortably close to the speed of light. The ensuing time dilation can then appear to freeze the action completely inside the ship. Thus, as long as we accept the fact that time intervals and distances are altered during a transformation from one frame to another (but always in such a way as to preserve the constancy of the speed of light), we must also accept the conclusion that the acceleration of one frame relative to another itself incurs an additional time dilation. Gravitational fields, therefore, slow down the passage of time as viewed from distant vantage points, and the retardation effect is greater the stronger the field.

Several other important consequences follow immediately from this effect, including the often-used (and observationally powerful) concept of a *gravitational redshift*. Sitting in a boat on a calm pond, we can create an outwardly moving wave crest by dropping a pebble in the water. Feeling ourselves adventurous, we decide to drop a pebble every 5 seconds. In this fashion, we generate a train of wave crests whose separation reflects the interval of time (5 seconds) between pebbles. Stretching the interval to 10 seconds causes the crests to rise less frequently, and therefore with greater relative separation. Shortening the interval has the opposite effect; it generates the waves more quickly and thereby reduces their relative separation. The distance between the crests is called the "wavelength," which evidently changes in inverse proportion to the frequency with which we drop the pebbles—the more frequently we drop them, the shorter the separation and hence the wavelength.

Now imagine that a beacon inside the spaceship is sending out pulses of light, which travel through the gravitational field out to large distances where we view this activity. The travelers measure a certain interval of time between pulses, but according to us, the interval is longer because of the dilation caused by the gravitational acceleration. So we infer a smaller pulse frequency (i.e., a smaller number of pulses per unit time) than that measured in the spaceship, and by analogy with the crests in the pond, we must therefore see a longer wavelength than that inferred by the travelers in the ship. Time slows down, the frequency decreases, and therefore the wavelength increases. This effect is known as a *redshift* because light of a given color is shifted toward the red end of the spectrum where the wavelength is longer. The stronger the gravitational field is, the more significant this effect becomes.

Several experiments have now shown that this phenomenon is real (and measurable!). In 1960, the physicists R. V. Pound and G. A. Rebka directed the 14,400 electron-volt gamma rays from radioactive iron on the ground floor of a 21.6-meter tower at Harvard University toward the top where similar iron nuclei were positioned to absorb them. This process is supposed to be reversible, in the sense that if an iron nucleus can emit a particular gamma ray, it ought to be able to absorb it, or another like it with very similar energy. Pound and Rebka reasoned that although Earth's gravity is weak, it is nonetheless not completely absent, and the difference in height between the base of the tower and its top should introduce a relative gravitational redshift. More specifically, gamma rays emitted on the ground should be redshifted by the time they reach the iron nuclei at the top, and this shift should be sufficient to quench their rate of re-absorption when they get there because of the mismatch in wavelengths.

Indeed, the results confirmed that the absorption rate was less efficient than normal. But when the iron nuclei at the top of the tower were made to move upward at just a particular speed, the gamma rays were then readily absorbed. The motion of the nuclei at the top was just enough to compensate for the gravitational redshift up the tower. This process works by analogy with the change in pitch (or frequency) that we sense when a train approaches and then recedes from us. Moving the absorber upward (away from the emitter on the ground) decreased the intrinsic frequency (i.e., "lowered the pitch") at which the gamma rays would be re-absorbed and therefore matched the redshifted wavelength coming up from the ground. This rather simple, yet elegant, experiment produced a fractional redshift within about 4 percent of the predicted value, and is considered to be one of the major achievements of twentieth-century physics.

In 1976, the National Aeronautics and Space Administration carried out its own version of this experiment using clocks instead of gamma rays, and a much bigger distance between the two events. Called the Gravity Probe A (or GP-A for short), the NASA mission carried a hydrogen maser clock to a height of 6,200 miles atop an expendable rocket; at this height, Earth's gravity is half as strong as it is at sea level. The acronym "maser" (microwave amplification by the stimulated emission of radiation) was coined by the Nobel Laureate Charles Townes and his collaborators in the mid-twentieth century, after they successfully demonstrated the principle of microwave amplification in their laboratory. The probe slowed, stopped for an instant at the apex, and then started falling back. During this brief moment, the scientists could measure slight differences between the flight clock and another one kept on the ground. The gravitationally induced time dilation agreed with the value expected from the Principle of Equivalence to within 0.007 percent accuracy.

3.6   BLACK HOLES AND THEIR EVENT HORIZONS

So the stage is now set for us to sweep confidently into the physical realm where gravity imprints its effects with the greatest emphasis. Brief as it was, Karl Schwarzschild's contribution to the theory of black holes was enduring because it provided a blueprint for mapping the behavior of spacetime in a way that still guides us in our exploration of this exotic landscape. In relativity, it's all about measuring distances and intervals of time, and comparing these between observers in different reference frames, such as the traveler in the aforementioned spaceship and someone sitting comfortably in an observatory far away from the source of gravity. Nature shows us that light possesses the highest attainable speed (even the effects of gravity cannot propagate faster, as we have seen) and that the distances and times we measure are altered between frames in such a way as to preserve this speed everywhere and always. Someday we'll understand why this happens, but for now we must be content with our achievement thus far in having produced a *metric* (i.e., a "ruler") to interpret these measurements and to be able to transform the inferred values from frame to frame.

In special relativity, which excludes the acceleration due to gravity, the metric is very simple to write down and to understand. Since light travels at 186,000 miles per second, it will have traveled a distance of 186,000 miles times the number of seconds that have elapsed in a given interval. So regardless of which observer is making the observation, his measured distance for a light pulse will be $d = ct$, in which c is the speed of light and t is the interval of time in seconds. To avoid complications with signs (since a light pulse can either be moving toward us or away from us, two cases with opposite signs in d), it is more convenient and conventional to write $d \times d = ct \times ct$, or $d^2 = (ct)^2$, meaning that the square of the distance traveled by light is the square of its speed times the square of the elapsed time. This relationship is the metric of special relativity, in the sense that no matter who is making the measurement, the observer will conclude that her value of d^2 for light is always given by $(ct)^2$.

Schwarzschild's contribution was to write down (by actually solving the gravitational field equations of Einstein and Hilbert in 1916) the analogous metric for situations in which gravity produces a local acceleration. As we have seen, the very act of letting time advance in a gravitational field produces a transformation into a frame moving even faster than the one before it. So now we cannot simply have $d^2 = (ct)^2$ since the final measurement for the determination of d occurs in a different frame than that of the initial measurement. A modification is needed in order to preserve the constancy of c; Schwarzschild concluded that this must be a factor that depends on the local value of the gravitational acceleration.

Schwarzschild's metric for the propagation of light in the case of a static, spherically symmetric source of gravity is $d^2/f = (ct)^2 f$, in which f is defined to be the factor $(1 - 2GM/c^2 r)$. Although the proof that this metric actually does satisfy the field equations requires some mathematical effort, understanding its physical meaning is not that difficult.[11] That's because the effects due to the gravitational acceleration enter in a very simple way. Recalling Newton's monumental effort in formulating the law of gravity, we recognize its presence (albeit in a modified form) within the factor f. In fact, the term $2GM/c^2 r$ (which contains the mass, M, of the gravitating object, a constant, G, that characterizes how strong the force of gravity is for a given inertial mass, and the radius, r, from its center) is the ratio of the escape speed to the speed of light, c, all squared.

Amazingly, we see that just as Michell and Laplace had predicted in the eighteenth century, something unusual does indeed occur when the magnitude of the escape velocity approaches the speed of light. Looking at this metric carefully, we notice that far away from the object, where the radius is large, the factor f approaches unity, which recovers the metric of special relativity (i.e., $d^2 = [ct]^2$). This limiting behavior is realized because the force of gravity decreases inversely as the square of the radius, so if the distance is large enough, the influence of *any* object becomes enfeebled. A far more interesting event occurs when we view what happens to light going the other way—toward the source of gravity. Then, the gravitational acceleration becomes so severe that time slows down considerably, indeed stopping altogether at a

---

[11] This form of the metric is correct as long as light is moving along a radius, r, that originates at the center of the mass concentration. So d here is a measure of the change in r only. If the ray of light is moving in any other direction, the correction due to the acceleration appearing on the left-hand side is then set equal to one. The correction associated with the passage of time (appearing on the right-hand side), however, is always present because of the effects we discussed in the previous section, having to do with the slowing down of time in a gravitational field.

critical distance known as the *Schwarzschild radius*, in honor of the person who first discovered it. It is evident from the metric that this occurs when f goes to zero, at $r = 2GM/c^2$. Using the language of Michell and Laplace, we would simply say that the Schwarzschild radius is the point at which the magnitude of the escape velocity equals the speed of light.

Objects that produce a spacetime with a Schwarzschild radius reachable by infalling matter (or light) are called *black holes*, for reasons explored in greater detail later. But to achieve a Schwarzschild radius takes some effort indeed. The Sun would reach that point only if it could contract so much that all of its mass then fits within a sphere no bigger than 3 kilometers in radius! As far as we know, the Sun will never do this because several methods of support will prevent it from collapsing to that unimaginably high density. At the galactic center, however, there does appear to be an object whose mass is 2.6 million Suns, all squeezed into an unusually small volume, with a Schwarzschild radius barely 10 times bigger than the Sun, fitting easily within Mercury's orbit! Even seasoned high-energy astrophysicists have difficulty grappling with the physical implications of such a catastrophic compression.

The Schwarzschild radius defines a special spherical surface dividing the exterior universe from the inaccessible, interior region of the black hole. Someone finding himself just outside of this surface can potentially move to larger radii and escape the dark entity's clutches without having to attain light speed to do so. But below this surface, even light cannot escape (see figure 3.8). A device falling toward this virtual membrane will appear to us to be slowing down progressively more and more as it approaches the Schwarzschild radius (due to the slowing down of time in the ever-strengthening gravity), while the radiation it emits becomes redder and redder, pushing toward the radio end of the spectrum and beyond. The gravitational redshift is most effective right at the surface, so our device emits its last detectable light signal upon reaching it. A consequence of this is that the interior region appears completely black—hence the designation of these objects as "black holes" and their "surface" as the *event horizon*.

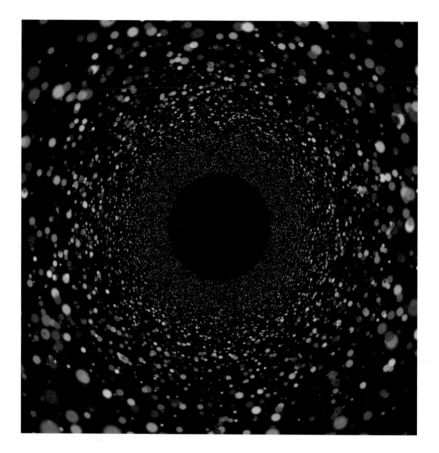

Figure 3.8 The event horizon is a virtual membrane that separates the external universe from the interior region, from which nothing can escape, not even light. As viewed from the outside, a completely isolated black hole would appear as a dark sphere, surrounded by the gravitationally lensed images of stars in the background (see also figure 3.6). The light from stars directly behind the black hole propagates right through the event horizon, and to the side of the ponderous object, the stars that would otherwise be present in the image are absent because their light is bent away from our line of sight by the strong gravity. This effect also produces the apparent high density of luminous objects around the black hole's rim. The colors are intended to illustrate variations in the brightness and surface temperature of the stars. (Simulation by B. Bromley at the University of Utah and the author)

These ideas about how things would appear near the black hole are based on our perspective from a distant vantage point. Curious as we are, it's natural to wonder what it would be like to actually fall toward the event horizon. Much of it depends on how close we can get, and that in turn depends on how massive, and therefore how big, the black hole is. Standing on the surface of the Earth, we feel immersed in a rather flat landscape (aside from the inconsequential mountains and valleys) because we're very small in comparison with the Earth's sphere. However, climbing to the top of the dome of the Adler planetarium in Chicago, say, we become very aware of the curvature of the surface below our feet. The smaller we are compared to the sphere, the more likely it is that we feel the same forces on one part of our body as we do everywhere else, because then it's like being on a flat surface. The Earth's tides result from the fact that the gravitational pull of the moon on our oceans is greater on the face closest to it than on the Earth's sides and back. In addition, the moon's gravity, pulling the oceans toward its center, actually squeezes the seawater onto the sides of the Earth, because from the moon's perspective, the Earth is very curved.

Something similar happens when we approach a condensed source of gravity, since all the forces on our body point toward its center. The smaller we are in comparison to the horizon's sphere, the less we feel the "tidal" effects stretching or squeezing our body. It turns out that for a black hole of eight solar masses (one that could form as part of the terminal evolution of a massive star), the tidal forces become too severe for us to survive once we approach within about 400 kilometers, well beyond the Schwarzschild radius of 24 kilometers for such an object. The horizon's sphere becomes large enough for us to avoid strong tidal effects once the black hole's mass exceeds 1,000 solar masses. So for the object at the galactic center, which contains millions of Suns, we could presumably fall through the event horizon without any significant disruption, except, of course, that the other matter falling in with us is hot and dense.

The Principle of Equivalence tells us that within our frame we would feel no effects due to gravity, since everything falling with us experiences exactly the same (mass-independent) acceleration. So for us, time would flow at the same rate as it would if we were very far from the black hole. Only the transformation between relatively moving frames and the acceleration that amplifies the consequent effects cause the distant observer to see a slowing down of activity as we approach the event horizon.

The Principle of Equivalence also affects what we see from the outside universe, since not all rays of light can reach us as we get closer to the Schwarzschild radius. Like everything else moving under the influence of a gravitational field, light bends toward the source of gravity. We can always see light approaching us from directly above (i.e., falling radially), but it gets harder to see light approaching from other angles as we get closer to the black hole. That's because light approaching us from other parts of the universe will curve progressively more and more as it descends deeper into the gravitational well, so that eventually all rays are falling along a radius. The image we have of the universe is compressed within a cone whose opening angle shrinks as we descend, eventually closing completely when we reach the horizon. Beyond this point, light can still reach us from outside, but nothing would be resolvable. We may get a sense of how severe these effects are by considering the simulated distortions to the background scene shown in figures 3.9 and 3.10.

No one knows for sure what exactly happens to matter once it crosses the event horizon, but an interesting suggestion has been proposed based on the form of the metric. What we wrote above is the metric applied to light. For a particle with nonzero mass, however, d^2 is never quite equal to $(ct)^2$, since its speed must always be less than c. Take a particle at rest in the laboratory. Time advances while its position does not. Thus, after a few seconds, d^2 is still zero, but $(ct)^2$ is definitely not. In special relativity, the metric for a particle with nonzero mass is written $s^2 = (ct)^2 - d^2$, which takes into account the fact that its speed is less than c, for which s^2 must then always be greater than zero. Remarkably, the fact that d and t change in just the correct way to preserve the constancy of the speed of light when going

Figure 3.9   Together with the panel shown in figure 3.10, this figure demonstrates the severe effects of light bending near a strong source of gravity. Shown here is a view over the sea and a spherical mass (in the form of a heavy soccer ball) hovering above it. Its gravitational force attracts the water below and creates a mound of fluid. The white bar points diagonally toward the backside, where it intersects a shiny, red object. Compare this with the image in the next figure to see how general relativity distorts this purely "Newtonian" view. (Ray-traced computer-graphic image courtesy of F. W. Hehl, R. A. Puntigam, and H. Ruder, and Springer-Verlag)

from one frame to another, also renders the value of s^2 immutable. Whatever value s^2 has for a given particle in one frame, it must have that identical value in every other frame. Not surprisingly, the Schwarzschild metric that incorporates the effects of gravity (though still ignoring motion other than radial) takes the form $s^2 = (ct)^2 f - d^2/f$.

If we believe, as Einstein did, that the rules of special relativity still apply even in the presence of a gravitational field, then we cannot escape the conclusion that s^2 must be preserved as we cross the event horizon. But wait! The factor f changes sign when we move from outside the black hole to the inside. It seems that time and distance must *reverse* their roles, since now s^2 can be due solely to a nonzero value of d^2 even if $(ct)^2$ itself is zero.

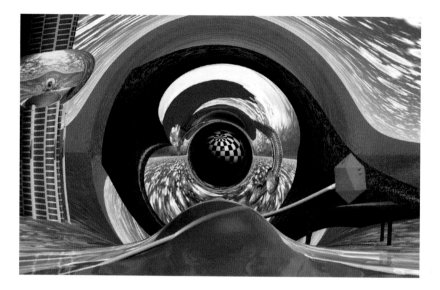

Figure 3.10 This panel shows the same scene as that in figure 3.9, though now with the inclusion of the correct light-bending effects. To avoid complicating the imagery too much, the small colored sphere in the previous panel was removed from the simulation shown here. Note also that light is redshifted progressively more and more as its source approaches the event horizon. This effect was not taken into account in producing these images so any light produced close to the massive sphere would in reality appear very red, or indeed, would be shifted so much that the central object should then appear black, as indicated in figure 3.8. (Ray-traced computer-graphic image courtesy of F. W. Hehl, R. A. Puntigam, and H. Ruder, and Springer-Verlag)

This exchange is why some people believe that once matter crosses the horizon, its motion toward the center is inexorable, just as on the outside we have no way of preventing Monday from turning into Tuesday. Speculating beyond this point is dangerous because general relativity is a classical theory. The past century has taught us well that only quantum mechanics can adequately explain the physics of phenomena on such compact scales as we would be likely to encounter approaching the so-called *singularity* at the center of the black hole. This name correctly conjures up imagery of infinite densities in dimensionless volumes, something quite alien to the concept of a quantum state.

Now that we've convinced ourselves that black holes are black, we might conclude that they are understandably extremely difficult to study. They have such immense gravity that nothing can escape their grasp, not even light—so how can we see something that does not emit or reflect any form of energy? We will address this question in the next chapter.

4

A STAR IN SAGITTARIUS

Nature must surely be a chiaroscurist, a grande dame of light and dark, of shadow and contrast. On its cosmic canvas, we can discern one of the greatest paradoxes in science—that black holes tend to be the brightest objects in the universe. And so it happens that images of the galactic center have thus far revealed not a dark spot, but rather a vibrant radiator like the central dot in figure 1.8. Sagittarius A*, it turns out, is truly a star in Sagittarius. Of course, this is due in part to our remoteness from that region, which makes it difficult for us to identify features as small as a Schwarzschild radius. Recalling the definition of this critical quantity $(2GM/c^2)$, we can estimate its value from the known mass of the black hole (2.6 million Suns), and find that it should correspond to roughly five solar diameters. That's not very big compared to the 28,000 light-year distance to the galactic center, and on an image such as figure 1.8, a Schwarzschild radius would be small indeed, at most one ten-thousandth the size of the central dot seen here. So it is not yet clear what exactly we have been looking at in the center of the Milky Way, though clearly its unusual character has something to do with the black hole.

In chapter 5, this situation will change considerably when we take it upon ourselves to push the bounds of technology further and aim for the event horizon itself, where even the exquisiteness of nature's art cannot conceal the darkest pit. But black holes so stress the space and time around them that regions *outside* the shadow of no return apparently emit radiation prodigiously.

This associated outpouring of radiation arises because Sagittarius A* does not exist in isolation from the rest of the universe, even though its interior is forever entombed within an unbreakable seal. Its very presence is felt with overwhelming force by any matter and radiation wandering haplessly nearby and, anyway, the advent of quantum mechanics allows for sufficient imprecision in the physical conditions just outside the event horizon that energy could still leak out slowly, though enduringly. The supermassive black hole at the galactic center not only influences the motion of stars in its vicinity, but it also shines by proxy, inducing the environment to radiate in its behalf, revealing its presence indirectly.

We are rather certain now that most of the radiation our detectors sense from Sagittarius A* is produced by the glowing, hot gas falling into oblivion past the event horizon, a concept that will occupy us for much of this chapter and the next. We are about to explore a region of space from which it would take us literally a matter of only seconds or minutes to plunge uncontrollably into the deepest abyss, and we will uncover one of the most exciting prospects since the early stages of the development of general relativity—that the black hole's shadow will soon be, after all, visible against the backdrop of luminous plasma. Black holes can "radiate" by several different mechanisms, however, not just by heating the infalling plasma and making it glow, so let us begin by examining each of these processes in turn to establish the overall context for this apparent paradox of nature.

4.1   HAWKING RADIATION

Even a perfectly isolated black hole (if such an object could exist) might reveal itself via a loophole created by the application of quantum mechanics, a theory that is known to be correct, if not complete. This subject continues to evolve and it would surprise no one to learn some day that these ideas are somehow flawed if the merger of relativity with quantum mechanics finally leads to predictions that diverge from what we know. The reason, as we noted earlier, is that general relativity is a classical theory, operating on the basis of precise measurements of physical quantities, such as distance and time. The very notion of defining an event horizon makes sense as long as we can precisely place this surface and particles around it at perfectly known locations. But along comes quantum mechanics, reminding us that there must always be some positional uncertainty, or an imprecision in energy and time, and that general relativity, like all other classical theories, must therefore acquire some minimal level of fuzziness.

Quantum mechanics argues that we can never be entirely sure of a particle's position or its energy because in order for us to even know of its existence, we must disturb it to sense its presence. Imagine being locked inside a totally darkened room, where no sound is permitted and all other forms of communication are excluded. How would you know if someone else is there with you? The only way you could determine if you were not alone is to wander around the enclosed space, touching, feeling anything that comes into contact with you. As far as you're concerned, you're alone until you hit someone else and prove that he is there.

It's actually more frustrating than this, however, since the hit only guides you into thinking that he's there, but it doesn't provide you with the solace of knowing exactly where he is. Even if you could somehow determine roughly where the hit took place, using an internal system of coordinates (say the approximate number of tiles from the west and north walls), you would still have no way of knowing exactly where he is after the bump. In our everyday lives, we acquire the illusion of precision only because the fuzziness induced by these uncertainties is very small, and our mind clings to the apparent clarity of the outside world as a convenient simplification of the way things really are. Certainly on a macroscopic scale, this fuzziness does not manifest itself readily, and our description of nature using exact positions and times is quite adequate for our need to interpret much of the activity in our environment. But on a microscopic scale this fuzziness is paramount, and nothing can happen without the consequences of this imprecision. We can build the most elaborate table-top experiment known to humanity, but unless a particle announces its presence by making contact with something inside it, we would never know it even entered the instrument.

This uncertainty is the reason why physicists are uncomfortable with the idea of a perfectly localized and sealed event horizon, since these notions make no reference to the fuzziness of quantum mechanics on the smallest scales. But physicists thrive on their despondency, which often leads to a scientific breakthrough just to assuage their deep anxiety. General relativity itself was the culmination of an 11-year struggle by Einstein to marry the special theory of relativity with gravity. And a phenomenon discovered in 1974 by Stephen Hawking may be the first step in the eventual resolution of the puzzle posed by the concept of an event horizon with quantum mechanical fuzziness.[12]

The name itself, *quantum* mechanics, reveals the essence of the physical description on a microscopic scale. Practically all of modern physics is based on our ability to determine the value of certain quantities, such as the intensity of light, in terms of their location within a prescribed volume. Quantum mechanics tells us that at this level all such entities are to be thought of as comprising tiny bundles (or quanta) of "something," which in the case of light are known as photons. In the appropriate terminology, one says that fluctuations in a field, say the gravitational field, are associated with the manifestation of these quanta, which can appear or vanish as the fluctuations grow or subside. The connection between these bundles and the fuzziness we introduced earlier is that their size, energy, and lifetime are directly related to the scale of the imprecision, that is, how fuzzy the measurements of position or energy turn out to be.

What we have just described is what happens when a field is actually turned on, for example, when we shine a light on something, meaning that a stream of photons is dispatched toward that object. However, it turns out that quanta such as photons may also bubble up spontaneously out of vacuum if an adequate

---

[12] Readers who would like to learn more about the technical aspects of this phenomenon, and the evaporation of black holes in general, will find the discussion in Thorne, Price, and Macdonald (1986) very helpful. See also Wald (1984).

source of energy lies nearby. But a crucial fact that we have gathered from the observed behavior of these fields is that when the bundles materialize spontaneously, they always do so in pairs, as if something must be split in order to create the fluctuation. So a quantum, or particle, with negative charge can only materialize if at the same time its counterpart, with positive charge, also comes into being. Given that every characteristic we can assign to this bundle must be matched by the opposite attributes of its partner particle, it makes sense then to talk of these as particles and antiparticles, or matter and antimatter. The annihilation we witnessed on a galactic scale in figure 1.14 is but one consequence of the coexistence—and eventual mutual destruction—of these particles and their opposites.

The phenomenon discovered by Hawking[13] is directly associated with this creation of quantum particles in vacuum due to fluctuations in the gravitational field of the black hole. Like a swarm of fireflies on a hot Georgia night, particles flash on and off just outside the event horizon. They live fleetingly and then annihilate with each other's counterpart to re-establish the vacuum after the fluctuation has subsided. But here's the catch. Let us recycle the gondola and water wave analogy and now think of the black hole as a giant gondola and the water as its gravitational field. The boat bobs up and down, creating outwardly undulating waves whose crest separation is in some way a measure of the gondola's size. Tiny boats produce short wavelength disturbances, while the waves from bigger ones undulate very slowly. By analogy, fluctuations in the gravitational field of the black hole have a wavelength commensurate with its size. So when these fluctuations manifest themselves as photons, or any other type of particle whose rest mass is small compared to the amplitude of the fluctuation, their wavelength, too, corresponds to the size of the black hole. Thus, in what appears to be yet another consequence of nature's chiaroscurist tendencies, the fleeting quanta produced beyond the event horizon of very massive black holes are much

---

[13]Some of Hawking's early discussion on this topic appeared in a paper published by *Nature* in 1974.

redder, and therefore of lower energy, than those associated with their smaller brethren.

The paired quanta produced in this fashion annihilate very quickly (in about one-millionth of a millionth of a millionth of 1 second) and, anyway, they're already outside the event horizon. True, but some pairs, argued Hawking, will have a member that dips below the membrane of no return, abandoning its partner to the whim of the outside universe. Without a partner to annihilate, the detached particle flees the black hole's sphere of influence and merges into the flux of escaping radiation headed for infinity. To us, this looks like the black hole is actually radiating, though the mechanism is clearly indirect. But one thing is clear, and this we can't forget: the source of energy for these fleeing particles is ultimately the black hole itself, and although we cannot claim that the radiation originated from within the event horizon, its energy surely did, and the dark object pays the price with a consequent decrease in its mass. Hawking had explored this process for the collapse of stars into black holes, reasoning that quantum particle creation would be significant during the rapid collapse due to the changing gravitational field. Expecting the emission of these particles to subside once the star had reached the singularity, he was surprised to find that instead the emission reaches a steady rate dependent on the photon's energy.

Treating the black hole as a quantum-mechanical object thus mitigates somewhat its effectiveness as a voracious devourer of everything around it, growing indefinitely until all food runs out. If this simple application of quantum mechanics survives the test of time, it appears that all black holes must evaporate eventually, though to us this might as well be an eternity. The Hawking radiation from a black hole with barely the mass of 30 Suns, let alone the 2.6 million solar-mass behemoth at the galactic center, has such a long wavelength, and is therefore so feeble, that it would take such an object 10^{61} times the current age of the universe to evaporate completely. That's 10 followed by 60 zeros! This process is hardly the stuff that makes Sagittarius A* one of the brightest radio sources in the sky, so its power must derive from something else.

4.2 ENERGY EXTRACTION ACCORDING TO PENROSE

Sir Roger Penrose[14] considered the possible consequences of the idea that real black holes are probably spinning because the material that forms them almost never falls in radially. When we open a door, we push against the handle far from the hinges and this action makes it turn with relative ease. It would be much more difficult to execute this simple task if instead we were to push on the door near the frame that supports it. In a similar fashion, material that falls through the event horizon away from the center causes the black hole to spin faster, or slower, depending on which sense of rotation it currently has. Over time, one might expect that if the rainfall of matter is random, the up-spins and down-spins would cancel, but the odds of getting a perfectly static object are rather small, since this would require an exquisite tuning not generally available to the vagaries of nature.

In 1963, almost half a century after the establishment of general relativity and Schwarzschild's war-front discovery of the first solution to Einstein's equations, a British-educated New Zealander by the name of Roy Kerr published a one and a half page article[15] announcing his discovery of the second solution, anticipated throughout those many years. Whereas Schwarzschild's solution described the fabric of spacetime surrounding a static black hole, Kerr's *metric* was an important generalization of this description to spinning objects, and given that most black holes in the universe are probably spinning, the new features introduced by Kerr's solution are rather consequential.

In fact, what has emerged is a most peculiar property of spinning black holes, whose current interpretation still may not be entirely convincing, though most physicists would be predisposed to accept it when the first signs of experimental verification appear. Whereas in a static black hole clearly nothing can sustain itself

---

[14]The original discussion appeared in Penrose (1969).

[15]This brief announcement appeared on pages 237 and 238 of the *Physical Review Letters*, Volume 11 (1963). Kerr followed this initial achievement with more elaborate calculations two years later.

against inexorable infall once it finds itself within a Schwarzschild radius of the singularity—you may recall from chapter 3 that the radial and time coordinates effectively change roles inside the event horizon so that progression toward the middle is akin to watching Monday turn into Tuesday—Kerr's solution shows that this property is not always true when the black hole is spinning.

It turns out that as the ponderous giant turns on its axis, its gravity also swirls the fabric of spacetime around it like water around a whirlpool. The interpretation of this phenomenon, known as "frame-dragging," is that even if a chunk of matter is not moving relative to space, the fact that space and time are themselves being forcefully dragged around the spin axis induces a motion in the object as well. So this matter *can* sustain itself against an inexorable infall, but only if it also moves sideways around the strong source of gravity. The bottom line is that the event horizon, that is, the surface of no return, must therefore be smaller when the black hole is spinning. The astrophysical implications are enormous; for one thing, matter can venture closer to the singularity of a spinning object, while preserving contact with our universe, than it can for a static one with the same mass. And in chapter 6 we will see that the existence of spinning black holes is probably what makes relativistic jets of plasma spew forth from the nuclei of many galaxies and quasars.

Evidence for the reality of frame-dragging may finally be trickling in, though this is not quite yet the compelling case that the experts are seeking. The *Rossi* X-ray Timing Explorer was a satellite flown by the National Aeronautics and Space Administration for the specific task of identifying the rapid variability of physical conditions of matter in orbit around compact objects, though not the supermassive variety like Sagittarius A*. It was looking instead at black holes only somewhat bigger than the Sun and bonded to stellar companions that feed them. Dancing in perpetual orbit around each other, these black holes and their partners draw ever closer with time, while the tenuous, frothy layers of the companions' surface shear off and swoosh toward the dark pits lying in wait several light-minutes or hours away.

Like the planets in our solar system, these wisps of gas orbit the black hole, though for them the struggle against gravity is pointless and they eventually fall in. In a handful of cases, the Rossi instrument detected the radiation eking out from the compressed plasma in its moribund struggle and saw this tragedy played out in real time. For one notable microquasar 10,000 light-years from Earth,[16] an unusual pattern in the detected radiation repeated itself some 450 times per second, a sure signature that it originated from incredibly hot gas dancing around the black hole in a very lively orbit well inside the region where a static spacetime metric would be capable of hosting it. It seems that the spacetime around this unusual object must be swirling around an axis in order for the hot gas to have orbited so close to the singularity without crossing the event horizon.

What if, wondered Penrose, this forced rotation of matter approaching a spinning black hole could be recycled into escaping energy? It would require some finely tuned dynamics, but the possibility does appear to be real, if not practical. In order for this process to work, a chunk of matter would need to fall inward and split into two pieces within the region where the spacetime is being dragged around the gravitational whirlpool. As long as one of these pieces ends up moving against the flow, which dooms it rather quickly with a plunge into the pit, the other piece gains enough energy to escape the trap and actually carry out more of it than the chunk had before entering the fray. Like Hawking radiation, this mechanism is an indirect way of extracting energy from the black hole. Both its spin and mass decrease to account for the energy loss. The dragging of the spacetime forces a corotation of matter entering this region and provides the added impetus to the forward-moving projectile.

This process can't go on indefinitely, however, since the black hole would grind to a halt and the so-called ergosphere (from the Greek word for *work*), where this action can occur, would

---

[16]The name microquasar is applied to a specific type of black hole whose mass is typically 10 to 20 Suns, with relativistic jets of plasma shrieking out in a perpendicular direction relative to the plane of matter in orbit around it. See Strohmayer (2001) for a technical account of this discovery.

vanish. As it turns out, the breakup of two particles inside the ergosphere must happen with a relative velocity of at least half the speed of light, which is difficult to do on an astrophysically large scale. Together with the delicate arrangement required for the energy extraction, this process therefore appears to be incapable of matching the much more powerful sources of radiant energy that we will explore next.

Incidentally, of the many peculiarities that a spinning Sagittarius A* would have, none would be more exotic and exciting than the prospect of entering it and advancing toward the middle. For a rotating black hole, the singularity is not a point, but a ring, and with just the correct, delicate balance of motion, one could conceivably plunge through its center and not get compressed catastrophically. Of course, the black hole's event horizon could never be re-scaled, so the rest of the universe would be lost forever. We can almost hear Professor Hardwigg muttering in the background that the intellectual satisfaction of traversing the black hole and reaching the true center of our galaxy may be worth it nonetheless.

4.3   COSMIC FIREWORKS

Well, some of us probably would not find this risk appealing, but many objects near Sagittarius A* do occasionally take the plunge. Sadly, for those that fail to enter cleanly through the event horizon, the price to pay is very dear. Sagittarius A East, the beautiful explosion of color in figures 1.4 and 1.6, is probably the splattered remains of an unfortunate star that didn't make it.

Unlike stars in the solar neighborhood, where the Sun and all its neighbors are granted a cordon sanitaire of several light-years, millions of their cousins at the heart of the Milky Way squeeze into a region no bigger than a light-year surrounding the deadliest predator in the galaxy. The most oppressed of them must not only contend with the unprecedentedly crowded field, but they do so at breakneck speeds approaching several thousand kilometers per second. Collisions in this swarm are not uncommon, and once every 10,000 years or so, the unluckiest star is knocked into a potentially catastrophic path toward the black hole. The rest of the sorry tale unfolding beyond this point constitutes the third radiative method by which Sagittarius A* may reveal its presence.

The errant star headed toward the middle must have a perfect aim for it to plunge directly into the black hole. Of course, this feat is difficult to execute, and is consequently rare. In most cases, its inward trek is like that of a comet knocked toward the Sun by one of the outer planets in our solar system. At first its plunge seems true, but as it approaches ever closer to the center, even a slight deviation from radial motion takes it off to the side, where it continues to swing around the rear. For the comet, at least, this brief encounter ends peacefully when the intruder recedes back to where it started, albeit lighter for its trouble, having lost its ice-cold surface layers to evaporation by the Sun.

That's where the comparison between the supermassive black hole and the Sun must end, however. Sagittarius A*'s gravity is so severe that no star could survive its swing around the back and live to tell the tale. Let us free our conscious thoughts for the moment, and imagine (unrealistically, of course) that a technological breakthrough is permitting us to construct a perfectly insulated sphere, able to withstand the heat and pressure inside the star. We're going to take the perilous ride wedged within the intruder's gaseous folds to see firsthand what misery awaits it.

We talked briefly in chapter 3 about the fact that when the black hole's mass exceeds about 1,000 Suns, the curvature of its event horizon is so slight compared to the size of our bodies that we would feel as if we were suspended above a plane in its proximity, in the same sense that the Earth's surface looks flat to us unless we view it from far away. In the case of Sagittarius A*, whose mass is significantly larger than 1,000 Suns, we could in principle pass through the membrane of no return without being torn to shreds, since our toes would feel about the same gravity as our nose. In other words, our whole body would be in "free-fall" in a space containing a relatively uniform gravitational field. So just as gravity disappeared for Einstein's unfortunate homeowner who fell from his roof, we find the black hole's force vanishing as we fall (never mind that an obliteration is likely to occur momentarily as we continue to head for the singularity in the middle). Cloistered within our protective sphere inside the star, we are therefore not likely to sense any immediate danger as it descends into the well.

But a star is 1 billion times bigger than we are, and for it the pull of Sagittarius A*'s gravity is far different on the face closest to the black hole than on its rear. In addition, just as the moon squeezes the Earth's oceans toward the middle of our planet, so too will Sagittarius A* compress the intruder's girth progressively with each passing moment of its descent. Astrophysicists have estimated that this tugging and pulling actually rips the star apart, stretching it along its path and squeezing it beyond recognition in the other directions.[17] This phenomenon, known whimsically as "spaghettification" (because of the long stretched out appearance of the object being ripped apart), continues until the star's compression is sufficient to tolerate further squeezing, which happens when its lateral dimension is about the size of our protective sphere. Looking outside, we would see ourselves immersed within a highly agitated plume of gas squeezed into a long, thin pancake.

This activity proceeds with increasing stress down to the deepest point in the star's descent, which doesn't even have to be the event horizon itself, as it turns out. The conditions at the galactic center are such that even a modest trespass to within 10 Schwarzschild radii of the black hole is sufficient to contort the doomed star irreversibly. To understand what is happening internally, imagine that a large number of springs are holding the star intact before this incident. The black hole's grip is so powerful that it cramps the star forcefully, breaking all the springs. The star continues to be squeezed far beyond the point where it would have any opportunity of recovering calmly once it left the black hole's clutches. This situation is like an exceedingly big version of a bicycle pump, in which compression heats up the gas, though here the black hole can increase the star's internal energy hundreds of times over what it had initially. And as the star exits the

---

[17]Many individuals have contributed to the general theory of "tidal" disruption, as this process is known, but the direct application to the galactic center is due primarily to Rees (1988) and Khokhlov and Melia (1996).

black hole's sphere of influence, the grip relaxes and no force is left to hold intact its highly agitated gases. It explodes with the power of 100 supernovas, each of which would otherwise be among the most energetic phenomena in the visible universe.

The ensuing fireworks fill the galaxy for hundreds of light-years across its center and they paint the sky with the schizophrenia of a fauvist's pallet (figure 1.5). Were we to survive the initial expansion within our sphere, this explosion would catapult us with speeds exceeding 5,000 kilometers per second into the galaxy's central bulge, where friction with the outside gases would raise the outside temperature to well over 1 million degrees, quickly burning the skin of our capsule, and thereby destroying us in a blaze of glorious sparkles. So much for the tourist-trade potential of this particular ride.

Not all of the unfortunate star's remains are ejected unceremoniously, however. Sensing the opportunity for a rare feast, the black hole clings onto some fraction[18] of the debris, which settles into a Saturnian-like ring of plasma orbiting the black hole's equator at near-light speed. For about a year thereafter, this disk glows with optical and ultraviolet light while neighboring rings of particles rub against each other to maintain the heat through friction. This sequence of steps also constitutes a futile exercise for the plasma, whose final plunge into the event horizon seals its tomb forever. Of course, the probability of actually witnessing this phenomenon live is rather small, given that the show is repeated only rarely, at most once every 10,000 years, and given that the fanfare is over in a wink of a cosmic eye. So all we have to work with is the graveyard of exploded stars long gone, but what a story they unfold!

4.4 SHAPE AND SIZE OF SAGITTARIUS A*

Astronomers have been watching the central beacon of our galaxy for several decades now, and though its output is anything but

---

[18]Recent calculations indicate that this fraction may be no larger than a quarter. See Shai, Livio, and Piran (2000).

constant, it appears that the light we see is from something other than the ghostly remains of stars that have fallen in. Perhaps some day, 10,000 years hence, our descendants will witness the once-in-a-multimillennium flare that signals another sacrificial destruction of a star in homage to the power of the supermassive icon of hidden worlds. They had better be alert, though, for a year later it will again be gone.

So what then are we seeing when we look at the (supposedly) dark-matter concentration at the galactic center? To answer this important question, which frames the pretext for the whole of chapter 5, we must do some careful sleuthing. Astrophysicists are a clever lot, and often lucky, squeezing valuable information from seemingly random data. Of course, some will say that the randomness is there for a valid reason!

To begin with, it would be extraordinarily helpful if we could see Sagittarius A*'s shape. Magnified views of the central region in figure 1.8 *do* show that the source is resolved, meaning that it's actually spread out—a smudge—not just a dot. However, it also turns out that we were just a tad over-zealous with our euphoric revelation in chapter 1 that the medium between the galactic center and us is completely transparent at radio wavelengths. But don't panic, since we're only talking about a temporary hitch. The problem is that the intervening gas is not uniform. It clumps in cloudlike fashion throughout the Milky Way, and this phenomenon can prevent us from seeing fine details in the radiating object.

To understand why this happens, put a ruler in a jar of water. Looking at it from outside the container, it appears that the stick is broken at the surface, with the inner piece pointing in some odd direction. This phenomenon, known as refraction, occurs because the speed of light in a medium such as water is lower than in vacuum. Think of a marathoner competing in the Olympics and attempting to break out of a crowded field of athletes after the start. She is able to run faster once she clears to the front (the equivalent of light propagating into vacuum) than when she is struggling to get through the pack (the thick medium). And this effect causes the image to bend. (There are other possible means of distortion, such as diffraction, which occurs when light is attempting to pass around an object.)

So the lack of uniformity in the medium between Sagittarius A* and us causes myriad tiny bends in the image as its light progresses from the source, through the clumpy medium, and into our detector. What would otherwise be a small source (perhaps even just a point), spreads out and becomes somewhat diffuse. We will recognize this effect operating in some of the images of chapter 5. Unfortunately, this spreading of the light also causes some loss in the identifiable structure, so the net result of our attempt to journey deep into the galactic center could have been a true disaster, a cruel trick of nature that led us to this point, only to let us flounder in frustration at the doorstep of an exciting discovery.

What saves us, and makes this exercise so interesting, is that the scatter-broadening of the image (as this spreading out is known) changes sharply with decreasing wavelength; in fact, it depends on the square of the crest separation. As we will see in the following chapter, nature has actually provided us with the opposite of a cruel joke—it has "arranged" for Sagittarius A*'s size to be just right for us to be able to see its intrinsic shape at a wavelength of about 1 millimeter or less, where the blurring from scatter-broadening is small. We'll soon learn that this wavelength also happens to be the one at which the event horizon starts to become visible through the foggy soup of glowing gas falling into the black hole. To see its shape, we must therefore continue the process of peering at the galactic center with progressively shorter wavelengths, until ultimately we acquire the undistorted image.

And here is where we approach the current technological limitations of radio astronomy. To get the type of image resolution we require to see fine structure with a scale of a few Schwarzschild radii, we must link together single-dish telescopes across the face of the Earth, from Hawaii in the Pacific Ocean to St. Croix in the Virgin Islands, creating a virtual baseline of 8,000 kilometers. That in itself is not the difficulty, since it has been done at wavelengths of 1 centimeter or longer.[19] But as these telescopes attempt to sense radiation with shorter and shorter wavelengths, several annoying problems start to creep in.

One of these hurdles simply has to do with the tolerance that can be maintained by the surface of each dish as the Earth turns and the Sun heats it unevenly. Expansion and contraction of the materials forming the saucer shape can cause deviations of millimeters or more, virtually annulling any hoped for sensitivity at a wavelength of this size. This range of wavelengths also happens to be where the Earth's atmosphere begins to absorb the radiation more effectively, reducing the intensity reaching the telescopes from space and therefore making images of faint patches of the sky even more elusive. There are more subtle technical issues involved as well, but all in all, few doubt that these problems will be overcome in the near future.

In the meantime, the best available images of Sagittarius A* have been formed at 7, and possibly also 3, millimeters, significantly better than the 2-centimeter scan used to produce figure 1.8. The curious thing is that Sagittarius A*'s size seems to change with the color of light it emits. The bluer the light is (meaning the shorter the wavelength, and therefore the higher the energy of the photons), the smaller it gets. So when radio astronomers image the black hole image at 3 millimeters, it is significantly smaller than when they capture its visage at 7 millimeters.

The best estimates we have to date reveal that Sagittarius A* is oblong in shape, with a major axis aligned in the upright position in figure 1.8, with an extent (at 7 millimeters) of about 72 Schwarzschild radii in the long direction and no more than about 20 in the short direction. With this diameter, Sagittarius A* could fit comfortably within the orbit of Mars in our solar system. Its size is less than half of this extent at 3 millimeters. Remember, however, that the scatter-broadening probably contributes to

---

[19]This technique is known as Very Long Baseline Interferometry, or VLBI for short. It links the signals received by about 10 individual telescopes or more, and with the very large effective diameter of the collection area, can provide an unprecedented level of detail in the high-resolution images.

Sagittarius A*'s apparent size substantially at these wavelengths, so we may not yet be seeing its actual shape. In addition, the scatter-broadening can be different in different directions, given that the gas in the galaxy is, after all, flattened into a pancake, so the oblong shape is surely due at least in part to this anisotropy.[20]

A very clever alternative method of determining Sagittarius A*'s size was used with great success in the early 1990s. The detective work goes something like this.[21] Imagine looking out at a light bulb and passing a small object between it and your eye. If you align this in just the right way, it will block some of the light as it passes directly in front of the bulb. Now try the experiment with objects of various sizes. What you'll notice is that the bigger the object is, the greater will be the variation of light intensity reaching you. In fact, you will see no direct light at all when the object is bigger than the bulb and right in front of it.

A similar experiment has been conducted with Sagittarius A* as the light source and the clumps of gas in the interstellar medium as the intervening objects. Astronomers have a pretty good idea of how big these cloud condensations are, from studying the light produced by pulsars and its temporal variations. These stars are spinning rapidly, sending out searchlight beams of radiation that flash past us as often as 30 times per second. No flickering has been seen in the galaxy's central beacon, meaning that Sagittarius A* must be bigger than the clumps, when viewed at a wavelength of around 1 millimeter. From this, we infer that its size must be at least 1.5 Schwarzschild radii at the shortest (meaning the bluest) wavelengths!

So there we have it—Sagittarius A* has a size that changes with color, ranging from something barely bigger than its Schwarzschild radius at 1 millimeter, to a value 50 times larger at 7 millimeters.

---

[20]For a recent technical review on this and related topics, see Melia and Falcke (2001).

[21]The report of this discovery was made by Carl Gwinn from the University of California at Santa Barbara and his associates (1991).

4.5   THE GLOW OF MATTER FALLING IN

Black holes are black and, if completely isolated, barely manage to reveal their presence with the feeble murmur of Hawking radiation. But Sagittarius A* is embedded within a matter-rich field from which it absorbs parcels of gas floating haplessly nearby. Rather than plunging quietly into oblivion, the captured plasma releases its ghost as a colorful glow before entering its tomb below the horizon.

We know that Sagittarius A*'s neighborhood is replete with particles because of images such as figure 4.1, taken recently with the Chandra X-ray satellite. Formerly known as the Advanced X-ray Astrophysics Facility, this state-of-the-art detector was renamed the Chandra X-ray Observatory in honor of the late Indian-American Nobel laureate, Subrahmanyan Chandrasekhar, and was launched in 1999 aboard the Space Shuttle. The nickname Chandra, which means "moon" or "luminous" in Sanskrit, is a very fitting name for this mission, recognizing Chandrasekhar's tireless devotion to the pursuit of truth. During his life, he made fundamental contributions to the theory of black holes and other phenomena that the Chandra X-ray Observatory is now studying, and he is widely regarded as one of the foremost astrophysicists of the twentieth century. Chandrasekhar won the Nobel prize in 1983 for his theoretical work on the physical processes that govern the structure and evolution of stars. With the ability to distinguish features barely one-twentieth of a light-year across at the distance to the galactic center, Chandra is providing X-ray images that are 50 times more detailed than those of any previous mission, and at more than 45 feet in length and weighing more than 50 tons, it is one of the largest objects ever placed in Earth orbit by the Space Shuttle.

Supplemented by the morphologically rich images of the galactic center produced with radio telescopes (see figures 1.7 and 1.9), the new X-ray views provide us with a wealth of information to complete the inventory of nonstellar matter in this region. Some of it is colder than the air in Antarctica, though the rest is hotter

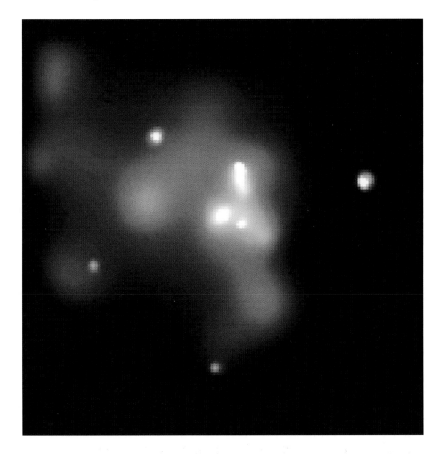

Figure 4.1 The Chandra X-ray telescope recently produced this image of the inner 10 light-year region of the galaxy, showing the distribution of hot, radiating gas. Sagittarius A* itself appears as the larger white dot near the middle of this view, slightly to the left of the smallest white dot. The gas surrounding Sagittarius A* glows in X-ray light because it has been heated to millions of degrees by shock waves from supernova explosions and colliding winds from young massive stars (see figure 4.2). Most of this plasma is expelled from the center of the galaxy, but some of it is trapped and absorbed by the black hole. This observation appears to be the first detection at X-ray energy of the supermassive object at the galactic center. (Image courtesy of F. Baganoff, G. Garmine, et al., and NASA/MIT/PSU. Copyright Massachusetts Institute of Technology)

than 1 million degrees Kelvin. The cooler material tends to condense into molecular clouds, whose internal density is 10,000 times higher than the density of the interstellar medium near the Sun. There are also some peculiarities having to do with the abundance of certain elements, which suggests a different history than what occurred out where we are in the spiral arms. A large fraction of the gas appears to contain recycled ash from the nuclear burning of stars from previous generations, but there is now evidence that some of it is drizzling into the heart of the Milky Way from virtually the beginning of time.

This inference is based on the detection of deuterium,[22] a heavy form of hydrogen, within 30 light-years of Sagittarius A*. Almost all of the deuterium in the universe was made within minutes of the Big Bang, which initiated the cosmic expansion over 13 billion years ago. Most of the elements in our bodies, particularly those heavier than carbon, were produced inside stars 1 to 2 billion years later, a process that also destroyed deuterium by burning it into helium. As such, the gas that has never resided within a galaxy should reflect the primordial constituents rather well. So the ratio of deuterium to hydrogen in gas left over from the Big Bang can be used not only as a measure of the density of ordinary matter in the early universe, but also as a tracer of untouched matter created at the dawn of time. The measured abundance of this element at the galactic center, representing about 1.7 parts deuterium for every million parts of hydrogen, is far greater than the ratio of one part per trillion expected if the deuterium was instead an unburned leftover from dead stars. At the galactic center, at least, the deuterium—and the gas in which it is contained—is raining down from intergalactic space and from the outside edges of the galaxy.

This primordial gas, mixed in with the recycled ash produced in stars, accounts for over 10 million Suns of matter in the extended region surrounding the galactic center. With its current diet, Sagittarius A* can thus continue to acquire mass for a long, long time. It cannot devour it all at once, however, because having

---

[22]This discovery was reported in the international journal *Nature* by Lubowich and his collaborators (2000).

this plasma settle into the middle is not as trivial as it may seem. For the same reason that the planets of our solar system don't just fall into the Sun, most of the gas skirts the galactic nucleus with sweeping orbits that keeps it temporarily out of harm's way. But bit by bit, a sliver here, a wisp there, its content dwindles as the inexorable gravitational pull from the center takes its toll, and streamers of plasma wend their way toward the middle. As we learned in chapter 2, this process appears to be the way in which the circumnuclear disk (figures 1.9 and 1.10) is replenished, providing a convenient launching point for the gas to plummet into the inner light-year region surrounding the black hole.

Once there, the trapped plasma furnishes a nursery that, every 10 million years or so, becomes ablaze with a new generation of bright, young stars, like those concentrated in the IRS 16 cluster shown in figures 1.10 and 1.11. And in the final twist of this contorted saga of how the gas eventually finds its way to the bottomless pit, it is the winds produced by these luminous objects that fill the space around Sagittarius A* with the supply of matter from which it feasts. These infant stars are so powerful and so tightly packed together that the stellar wind each produces collides with that from all the others, forming a halo of 60-million-degree gas. Figure 4.2 shows the resulting cauldron of seething plasma produced in this fashion within half a light-year of the supermassive black hole. This gas glows brightly in X-rays, filling the central cavity with the energetic radiation captured by Chandra in figure 4.1.

Playing Russian roulette, the clumps of gas in this tessellated pattern bob and weave, forward and backward, until one or more venture to within a twentieth of a light-year of the black hole, and then their fate is sealed. For a black hole with a mass of 2.6 million Suns, this is the point of no return—the threshold across which an endless fall to the central singularity is unavoidable. But like the wayward stars that on occasion get knocked toward the black hole in comet-like fashion, these clumps of gas hardly ever fall directly in, since even slight deviations from radial motion will have them swinging to the side and around the back. They merge into the rest of the gas that has settled into a disk sweeping

Figure 4.2 Modern hydrodynamics codes have the capability of simulating the behavior of gases in three dimensions, subject to many physical influences, such as the gravitational pull of a central mass condensation, magnetic fields, and the pressure due to radiation. This image is a "snapshot" of the column density (i.e., the gas density integrated along the line of sight) taken at a point in the calculation when the gas distribution had reached stationary equilibrium. Sagittarius A* is in the middle of this false-color image, and the dimensions are approximately 0.5 light-year on each side. Some 15 to 20 stars surrounding the black hole produce winds that collide and form this tessellated pattern of gas condensations, some of which are captured by the black hole and subsequently accrete toward it. Several of the wind-producing stars are visible to the right of the image. The color scale is logarithmic, with a dynamic range of about 100,000 from maximum to minimum density. The overall dimension of the simulation shown here corresponds to the central one-twentieth of the region photographed by Chandra in figure 4.1. (Produced by R. Coker at the University of Leeds and the author)

around the central object within its clutches of strong gravity, and spend their final weeks in our universe drifting inward while the unimaginably stressed conditions threaten to break the laws of physics as we know them.

You'd think that nature would be kinder to matter that has completed a 13-billion-year journey from its creation in the Big Bang to its fated endpoint hovering above the precipice of oblivion. It's ironic that these particles, born in ultimate violence, must end their life in our universe under conditions not seen anywhere or anytime since that pivotal event that initiated all of space and time. It's difficult for us to acquire a visceral feeling for how punishing these final days are on the infalling gas, but let us try to describe the events in the following manner. In its fatal progress from the point of capture by the black hole to its eventual status just before it crosses the event horizon, the plasma is compressed 1 million billion times. And as it crosses from one level of high density to the next during its descent, its temperature correspondingly rises. By the time the gas has reached the final layer just before disappearing through the membrane of no return, its temperature will have risen to 100 billion degrees or more—that's 10 million times hotter than the surface of the Sun!

It is easy to understand why we should therefore see a glow surrounding Sagittarius A*, and why its size changes with color. As the temperature of the gas rises, the closer it gets to the middle, the more energetic is the radiation it produces, meaning that the gas emits bluer and bluer light as it approaches its tomb. Based on the measurements we discussed earlier in this chapter, it appears that the light produced during the final moments has a wavelength of around 1 millimeter or less. This is the magical sprinkle of knowledge we need to craft our final approach to the black hole, which we complete in the following chapter.

5

THE EVENT HORIZON

In a series of influential papers published in the early 1970s, a pair of physicists from Yale University and the University of Washington in Seattle[23] posed what at that time appeared to be a mostly esoteric question, namely, "What would a black hole look like if we had the technology to actually be able to see it?" Of course, back then, the Hubble Space Telescope was only a twinkle in the eye of Congress, and X-ray astronomy was barely getting off the ground with the launch of single-stage rockets propelling crude devices into Earth's upper atmosphere. Radio astronomers would not be detecting a mysterious bright source at the galactic center for another year or two, and astrophysicists were still puzzling over the meaning of a claim that gravitational waves had been identified from the nucleus of our galaxy.[24] All in all, the prospects for actually seeing the image of a black hole that they calculated seemed rather thin. They reasoned that such a calculation was still scientifically useful, however, since knowledge of what the black hole looks like at the source could provide us with an indication of how

---

[23] J. M. Bardeen and C. T. Cunningham developed the first detailed image of an object whose light is gravitationally focused by a black hole between it and the observer, and reported these results in a paper published by the *Astrophysical Journal* in 1973. Many other researchers have followed in their footsteps since then, adding important new features and adopting more realistic environments for their calculations.

[24] See Weber (1969). Unfortunately, others have tried to confirm these early signs of a detection, using even more sophisticated devices, and never succeeded.

the light produced in its vicinity might vary with time, giving us the opportunity of monitoring suspected black hole candidates in an attempt to identify their predicted signature. "Observable effects produced near the event horizon," they wrote, "are particularly interesting, since they test strong-field predictions of general relativity."

And yet, it wasn't until about two decades later that this interesting idea found a pertinent application to Sagittarius A*. The first concern was how effects such as light-bending and gravitational redshift (see chapter 3) would restrict the type of radiation that could escape the supermassive black hole and reach our instruments.[25] Then, in 1994, a former U.S. marine and nightclub performer from Arkansas by the name of Jack Hollywood—a nom de plume derived from his previous persona—decided after several attempts at establishing alternative careers that his true calling was physics and enrolled into a full-time graduate program at the University of Arizona. He was a brilliant student, cutting through course work and engaging himself with research at an enviable pace. Almost immediately, he decided that general relativity was what he wanted to learn above all else, and moved effortlessly into a study of the effects of strong gravity associated with all aspects of Sagittarius A*, including its appearance.

The growing sentiment at that time was that although nearby stars could reveal the massive nature of this unusual object via their motion (see chapter 2), they were nonetheless still 50,000 Schwarzschild radii away from the center. This distance is far closer to the central black hole than is seen in any other comparably massive object in the nuclei of other galaxies, but it is nonetheless still far outside the region of strong gravity. To many astrophysicists, arguments concerning the nature of Sagittarius A* based solely on dynamical measurements are indirect at best, and can never fully satisfy the need for a compelling "proof" that its character is forged in the crucible of general relativity. A true

---

[25] Read about these early ideas in Melia (1992), and Falcke et al. (1993). See also the relevant subsequent work by Duschl and Lesch (1994), and Narayan, Yi, and Mahadevan (1995).

appreciation of its black-hole status will come only when we can "view it" directly.

The soon-to-be-minted Dr. Hollywood and his collaborators published several papers[26] outlining how general relativity ought to influence every aspect of what we can see from Sagittarius A*: its radiance, how this luminosity should vary with time, and the appearance of its environment if we had the technology to resolve it. What was lacking was a full appreciation of how quickly telescopes would improve, particularly those that detect radiation at millimeter wavelengths. In the next section, we will see why light in this portion of the spectrum is so crucial to understanding the nature of Sagittarius A*. Adding this last piece of the puzzle has brought us to where we are today.

5.1   THE ENVIRONMENT NEAR THE PRECIPICE

Having seen the gondolas in Venice and having learned all the physics they have to teach us with regard to waves, dust, and supermassive black holes, we next head out west to visit *Il Duomo* in Milan. But as we approach the piazza out front we realize that an enormous crowd has gathered to participate in a very colorful festival. Each person there is apparently dressed in one of three colors—red, yellow, or blue—and we can estimate by simple inspection that there must be 10 yellow people for every blue person, and 10 red for every yellow (never mind what this means— it's just an analogy!). To get to the entrance of the cathedral, we must thread our way through this throng of humanity, and since we don't want to lose sight of our target, we keep an eye ahead to align our internal compass. We're very observant and we notice that most of the people whose path we cross are dressed in red. In fact, given the over-abundance of this color, the piazza looks like a sea of red sprinkled with some yellow, and a few spots of blue

---

[26]The reader interested in sampling the literature on these developments can find the relevant discussion in Hollywood and Melia (1995), Hollywood et al. (1995), and Hollywood and Melia (1997).

dispersed thinly from corner to corner. Sometimes, we encounter a person in yellow, but only rarely do we bump into a blue one. This differentiation is hardly surprising, we think, given the ratio of colors. To minimize the confusion, we train our eyes to see just red, but this color is not ideal since our line of sight is cluttered with it, and we quickly abandon it. Next, we try yellow, hoping that our mind's filters can block out everything else. This color is much better, since now there are far fewer individuals competing for our attention as we focus on the entrance. But we might as well go one step further and filter out all but the color blue. This choice turns out to be the best, since now there are hardly any people between our target and us causing our eyes to defocus, and the doors to the cathedral's entrance project out vividly in the distance.

The medium, which in this case is the crowd, becomes more "transparent" as our mind's filter rotates from red, to yellow, to blue. Something analogous to this occurs within the excruciatingly hot gas churning above the event horizon in Sagittarius A*. Looking through this plasma, we can peer to greater depths depending on how we adjust the frequency (i.e., the color) of the radiation that we sense. Most of the light escaping from this environment is produced by a process not unlike that associated with the auroras, though enormously larger in scale. The aurora borealis (also known as the northern lights), and the aurora australis (southern lights), are beautiful, dynamic, luminous displays seen in the night sky near Earth's poles. These phenomena encircle the entire polar regions, but when viewed from the ground, only a small portion is visible, usually as a curtain or ribbon of thin, glowing gas. This activity begins when a flood of electrified particles, moving at a million miles per hour, stream out from the Sun and pass by the Earth. Shielded from the full blast of their destructive power by Earth's magnetosphere—a large bubble of magnetic field that deflects the solar wind—our planet's surface is immune from the direct impact of these very energetic projectiles. But some of them end up spiraling about this field, which can be thought of as a bundle of tight strings wound around Earth's atmosphere, and bounce back and forth from pole to pole. These particles gradually lose their energy by emitting light in various colors, some of it as a result of being flung about by the magnetic field (this is known as synchrotron radiation) and the rest by colliding with atoms in the air above us. We don't see much of the former, since that light tends to have a wavelength in the radio portion of the spectrum, but we are generally quite aware of its presence because of the interference it causes with wireless communication.

Some 5 to 10 Schwarzschild radii above the black hole, the infalling gas begins its final approach, spiraling inward with alarming speed and compression. The pulling and squeezing stretch the tendons of the magnetic field linking the parcels of hot plasma, forcing the highly agitated particles into a frenzy of collisions and gyrations. And on a scale trillions of times more intense than in Earth's magnetosphere, they stage a grand aurora borealis as a fitting curtain call before their exit from our universe and their irreversible entry into the interior of Sagittarius A*.

But rather than creating a nuisance with our communications, the radio emission from these particles is the means with which the doomed plasma reaches out in its final desperate attempt to let us acknowledge it once existed. Using the colorful Milanese piazza analogy, we would say that the bluest light—that with the highest frequency—is produced in the deepest regions where the conditions are the most extreme, that is, where the gas is the hottest, its density is the greatest, and the magnetic field is the most intense. At the same time, since there is far less plasma producing light with this wavelength compared to the rest of the radiation, the medium becomes more transparent as we train our radio telescopes to look at Sagittarius A* at progressively shorter wavelengths. So just as we could see the cathedral's entrance clearly by focusing on the blue light only, we can actually see all the way through the infalling plasma when we focus on the radiation from Sagittarius A* that began its escape a breath above the precipice. This thinning out begins to take effect at a wavelength of about 1 millimeter.

Although it was realized in the early 1990s that the medium surrounding Sagittarius A* should become transparent with an appropriate tuning of our radio telescopes, it was not known with certainty until recently that the transparency ought to be associated with plasma this close to the event horizon. And that makes all the difference. This concept is so important that it merits paraphrasing: by looking at the supermassive black hole at the galactic center with progressively bluer light, we see deeper into the abyss as the medium becomes more transparent. At the point when we can view the medium all the way through to the other side, we are actually sensing radiation produced primarily by the infalling gas on the edge of its ultimate plunge—we are therefore *seeing* the horizon itself, or more accurately, the "shadow" of the black hole (see figure 5.1).

5.2   HOW THE DARK SHADOW FORMS

Many astronomers expect that within the next 10 to 15 years, an announcement will be made to report a successful imaging of Sagittarius A* showing many of the features evident in figure 5.1. This event will bring a certain measure of closure to the often fitful, century-long search for the viability of general relativity as a description of very strong gravity. This is not to say that our understanding of gravity will by then be complete. On the contrary, there is much to do, particularly since the quantum version of this force is still a concept in its infancy. What this discovery will do, however, is confirm the bizarre prediction that a black hole should have an event horizon and that its internal infrastructure is completely divorced from the universe we know.

The appearance of the shadow that they will see is a natural consequence of Sagittarius A*'s powerful influence on the fabric of spacetime. If we can see all the way through the veil-thin plasma with its gossamer glow of twilight, we can't help but peer directly at the dark pit when we orient our view toward the middle. Light emitted by the gas directly behind it won't reach us because along the way it passes through a region where gravity is so strong that it traps the radiation permanently within the horizon. This

Figure 5.1 This image is a simulation of the thin, radiating gas surrounding the black hole at the galactic center. The black hole's "shadow," slightly bigger than the event horizon, betrays its presence. Because of its strong gravitational field, light from the infalling gas behind Sagittarius A* does not make it directly to us but is rather bent away or simply absorbed into the darkness. This effect manifests itself as the absence of radiation—a shadow— at the edge of Sagittarius A*'s event horizon. The image shows a general relativistic ray-tracing simulation for the appearance of the black hole if there were no intervening dust and gas between us and the galactic center. The shadow is predicted to be about 30 microarcseconds across—a diameter that should be resolved within the next few years by currently evolving very long baseline interferometry at sub-millimeter wavelengths. (From Falcke, Melia, and Agol 2000)

situation is analogous to a solar eclipse, in which the moon passes between the source of light—the Sun—and us. All of the Sun's rays propagating in our direction within the moon's orb are intercepted by the lunar surface and we see darkness—a shadow—where the pock-faced interloper blocks the view. To either side of the intervening object, however, the light passes through unobstructed, and we can see it arriving from all depths in front, to the side, and beyond the intruder. For example, even during a solar eclipse, we can still see the Sun's frothy expulsion of tenuous coronal plasma expanding beyond the moon's silhouette. For the same reason, we have no problem viewing the radiant gas to either side of Sagittarius A*.[27]

But there is a very important difference between the moon's capricious attenuation of the Sun's rays and the black hole's absorption of the light radiated by the plasma behind it. The moon's gravity is so weak, comparatively speaking, that light rays passing by it are effectively moving on straight lines; this is certainly not the case for Sagittarius A*. At the galactic center, the gravitational pull is so strong that even light, moving at the highest attainable speed, is bent severely. The closer a ray approaches the event horizon, the greater is the curvature of its path, so that beams not directly incident on the black hole's "surface" (defined to be the sphere with a Schwarzschild radius) are nonetheless still diverted away from us.[28] (We will see that these photons sometimes orbit the black hole and escape to infinity, producing mercurial patches and rings of secondary light.) The shadow of the black hole is therefore bigger than its event horizon because in addition to removing the light approaching it directly from behind, the black hole also flexes the rays grazing its surface into directions out of our line of sight.

Physicists readily calculate the shadow's appearance using a technique known as ray-tracing, which can produce black-hole images, such as those in figures 3.7 and 3.8. In the same way that

---

[27] The haunting image of the black hole's shadow in figure 5.1 was rendered by Eric Agol, who is currently a Chandra Fellow at the California Institute of Technology in Pasadena.

[28] See Falcke, Melia, and Agol (2000) for a full discussion.

we can calculate the trajectory of a baseball hurled from left field toward home plate, we can track the motion of a single ray from its point of emission near Sagittarius A*, outward through the intense gravitational field, and into free space where sentient beings like us can detect it and marvel at nature's wondrous constitution. Starting with Bardeen and Cunningham (1973), and followed by others in the intervening decades, researchers have used these simulations to show us that a black hole's shadow ought to have a diameter corresponding to about five times the Schwarzschild radius. The most recent application of this idea to Sagittarius A* (see figure 5.2) presents us with the tantalizing realization that a hole of this size in the otherwise glowing plasma is but a factor of only 2 smaller than the smallest feature now resolvable by radio interferometry. But before we survey the status of observations planned to test these predictions, let us complete our discovery tour of the black hole's shadow by examining another crucial property that merits our attention.

This new feature has to do with how light "wiggles." The wave description is generally quite powerful, but it often disguises the fact that what is waving has many directions in which it can crest and wane. Think of a screen onto which we project an image from a device some distance away. As the light propagates from the source to the screen, the radiation field can wave up and down, left and right, or anywhere in between. In other words, the wiggling occurs in any direction within a two-dimensional plane, perpendicular to the third spatial dimension pointing from the projector to the screen. Most sources of light pump out radiation that wiggles in all the possible directions, so even if we could in principle distinguish between photons that wiggle up and down from those that wiggle left and right, our screen would still be a completely randomized jumble and we would just see a filled region with light wiggling everywhere. However, some sources of radiation do favor a particular direction of oscillation, and some wiggling directions are treated preferentially when randomly waving light encounters matter. Looking around us, some of the light we sense is therefore randomized, while the rest is vibrating in preferred directions.

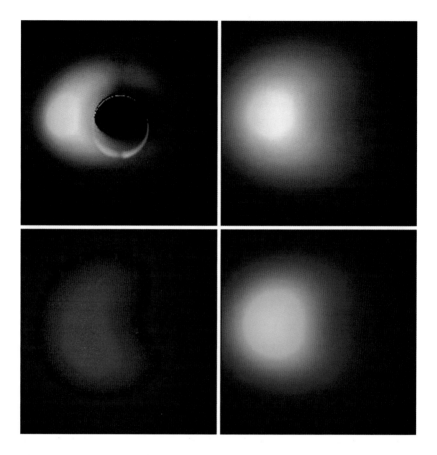

Figure 5.2 A more refined study of the gas falling toward the event horizon in Sagittarius A* suggests that it spirals to form a flattened disklike profile in orbit around the black hole. However, this does not affect the expected size of the "shadow." In addition, when the gas orbits in such a highly ordered sense (shown for the case of a counterclockwise rotation), the emitted radiation may be viewed with filters that reveal its direction of oscillation, that is, its *polarization.* In these simulated images (calculated at 1.5 millimeters), red corresponds to radiation oscillating in the vertical direction, whereas blue shows the brightness map of radiation oscillating in the horizontal direction. The top-left panel shows the black hole as it would appear to an unobstructed detector. The ringlike feature is produced by radiation that orbits once around the black hole before escaping. The top-right panel shows the image after scattering of the radiation by interstellar gas and dust. The bottom-left panel is the same as the top-right one, except showing only the vertically polarized radiation, and the bottom-right panel shows only the horizontally polarized radiation. (From Bromley, Melia, and Liu 2001)

We may not always recognize this phenomenon, but most of us have at some time or other encountered its practical consequences. The manner in which polarized sunglasses work is particularly attuned to how light vibrates. The direction of wiggling is known as the polarization of the light, and the material in these sunglasses has the special property of transmitting only one sense of polarization. It turns out that the glare we see from the reflection of the Sun's light by cars and windows and many other shiny surfaces is composed of photons with predominantly one direction of light vibration. So if the material in the lens is oriented just right, the glare is reduced because the light cannot pass through. If you've ever tried crossing two lenses with perpendicular polarizations, you'll remember that all of the light incident from the other side is extinguished and the lenses look dark. This darkening is due to the fact that the first lens blocks out everything except one direction of light vibration, which is itself then attenuated by the second lens that is letting through only the opposite polarization. As you rotate one relative to the other, however, the light begins to filter through, reaching maximum intensity as the lenses become parallel.

The same principle is at work when we don glasses in a movie theater to watch a film shot in three dimensions. With our naked eyes, we actually see on the screen a blending of light with two distinct directions of polarization, and since our eyes can sense both of them, the 3D image looks blurred because the two views are slightly displaced relative to each other. Looking through the glasses, however, one eye sees only the light with the up and down wiggling, whereas the other sees the complementary radiation waving sideways, and together with the displacement of the images on the screen, this creates the illusion of depth.

The radio photons from the aurora borealis are also polarized, because the particles that produce this radiation are not free to move in all directions. Instead, they're compelled to flow along magnetic field lines, like a tram on its tracks, so when they fling off their radiation, the wiggling direction is preset by their ordered motion. It's a little bit like synchronized swimming, with all the performers acting in unison, creating a grand display of harmonious motion. Because these particles bounce back and forth along the same direction, their collective radiation tends to have the same sense of vibration, induced by the underlying magnetic field.

The hot plasma flowing around Sagittarius A*'s girth generates its own powerful version of the aurora borealis, with a magnetic field combed out along concentric circular tracks. We see one sense of polarization when we look at the front and rear of the black hole, and quite a different polarization direction when we sense the light coming to us from its sides. So if we could get sufficiently close to it and put on our 3D cinema glasses, one eye would see light from the front and rear, while the other would see only the light produced on its sides. What if our telescopes could form not only a total image of the glowing gas and the black hole's shadow as in figure 5.1, but go one step further and image each of these polarizations separately?

This procedure is not as difficult as it may sound, since the technology to build a polarimeter—basically a large version of polarized sunglasses—does exist and has been employed successfully with a broad range of radio wavelengths. With such a device, millimeter telescopes will be able to see the views displayed with the impressionistic artistry[29] shown in the panels of figures 5.2, 5.3, and 5.4, where the light reaching us from Sagittarius A* wiggling up and down is shown in red and that vibrating left and right is rendered in blue. In these images, the gas is orbiting the black hole in a counterclockwise direction, so that the radiating particles are approaching us on the left and receding to the right, which accounts for the crescent-shaped appearance of these maps.

Each of these figures is a prediction of what we expect to see with the new generation of millimeter telescopes, taking into account the various effects we discussed previously, that is, the defocusing caused by refraction and diffraction in the interstellar medium between the galactic center and us—a problem we encountered in our discussion of Sagittarius A*'s size and shape—and the limitations on our attainable resolution given the finite

---

[29]Courtesy of Benjamin Bromley at the University of Utah.

size of the radio dishes and their separation across the face of the Earth. In each of these figures, the top-left panel shows the ideal image of Sagittarius A* if none of these factors were at play. The actual map that radio astronomers will make is shown in the adjacent panel on the right. The bottom two panels reveal the images we will see by using the polarized filters with the detected millimeter radiation, showing just the light wiggling up and down (in red) on the left and the radiation vibrating left and right (blue) on the right.

In the sequence of sampled wavelengths (from 1.5 millimeters in figure 5.2 to 0.67 millimeter in figure 5.4), we can see clearly how the black hole's appearance sharpens up dramatically even when the defocusing effects due to scattering and finite telescope resolution are taken into account. Millimeter astronomy should therefore be able to provide us with a viable image of the black hole's shadow *and* the moribund plasma flowing at near lightspeed toward its permanent interment. For these reasons, astrophysicists now sense that we are sitting at a landmark juncture in science, where the technology of the near future can produce these images at the actual wavelength where Sagittarius A* becomes transparent.

Unfortunately, one other jewel in the treasure box of discovery won't be accessible anytime soon. In the top-left panel in figures 5.2, 5.3, and 5.4 there also appears an additional, bright ring of light, floating above the shadow like the real image in front of a concave mirror. This is the light that failed in its first attempt to escape from the black hole's clutches, but managed to swing around the event horizon once, catapulting itself to infinity on the return trip. Only a very tiny range of orbits will permit this to happen, so the ring is necessarily razor thin.

Extrapolating this further, we expect there to be an even thinner and fainter ring associated with light that was similarly bound to the black hole at the beginning, but managed to escape after two orbits. Of course, the likelihood of this happening decreases rapidly with each successive loop around the horizon, so seeing higher-order rings is very difficult indeed. Even the first ring gets

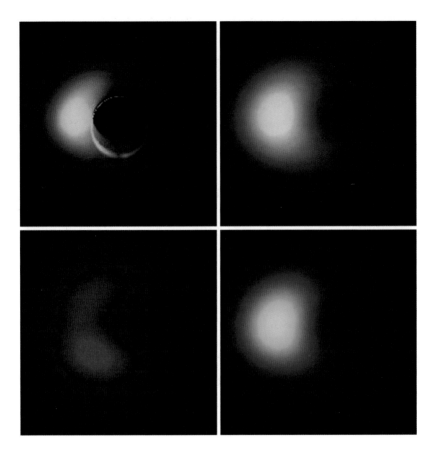

Figure 5.3 These images have the same meaning as those figure 5.2, except here they are shown for a wavelength of 1 millimeter. At this wavelength, the obscuring effects of scattering by the interstellar gas and dust have begun to subside, and we can see the shadow and orbiting gas more clearly than at 1.5 millimeters. The clarity continues to improve as the wavelength of the radiation decreases further (see figure 5.4), though this effect is somewhat mitigated by the fact that the source also gets fainter. (From Bromley, Melia, and Liu 2001)

washed out by the interstellar scattering and defocusing, dashing our hopes of using this purely general relativistic effect to perform additional tests on the behavior of spacetime with strong gravity. No doubt, this hurdle will itself be overcome in due course.

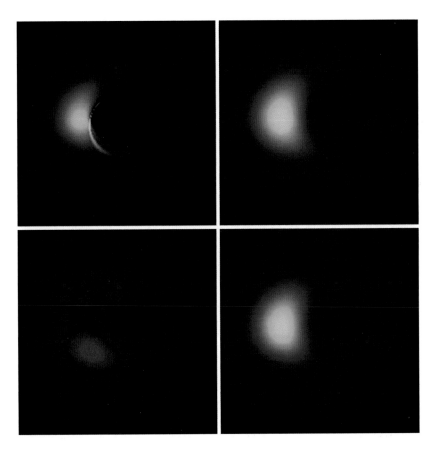

Figure 5.4 These images are the same as those of figure 5.2, except here they are shown for a wavelength of 0.67 millimeter. By the time we reach this wavelength, the effects of interstellar scattering have almost subsided completely. It would be desirable to view Sagittarius A* at even shorter wavelengths, but it appears that the infalling gas becomes too faint for us to see it much below the wavelength assumed here. (From Bromley, Melia, and Liu 2001)

5.3 AN IMAGE OF THE BLACK HOLE WITHIN THIS DECADE

The development of technology required to complete our exploration of the galactic center has already been underway for several years. We touched on some of the hurdles that need to be overcome in chapter 4, when our primary concern was how to determine Sagittarius A*'s shape and size at radio wavelengths, not just at 1 millimeter, where its radiating gas becomes transparent. Their impressive stature notwithstanding, existing radio telescopes are not all employable at the shorter wavelengths because of the difficulty in maintaining sufficient structural integrity to provide a pure millimeter or sub-millimeter signal. So a major problem with conducting worldwide coordinated observations at these wavelengths is simply the paucity of appropriate sites.

The idea for developing a coordinated global network of millimeter telescopes, which has come to be known as CMVA—an acronym derived from Coordinated Millimeter VLBI Array—actually goes back to the mid 1990s, when members of the Haystack Observatory in Massachusetts developed plans to create the network for initial observations at 3 millimeters, and additional experimental observations at 1.3 millimeters. Since then, the goal of the CMVA has been to continually break new technical ground for later exploitation at progressively shorter wavelengths.

Given that this is still a relatively new field, the group of astronomers and engineers working on millimeter VLBI (which in turn stands for very long baseline interferometry, a process discussed in chapter 4) is understandably rather small, numbering somewhere in the range of 50 to 100. Since 1996, new VLBI experiments have been performed at 3, 2, and 1 millimeters. At 3 millimeters, up to 12 stations around the world have been able to participate in global VLBI sessions, organized through the CMVA, twice a year. Although unfavorable weather conditions and technical problems at some sites sometimes affect them, these campaigns are generally successful and provide good observations of compact emitting regions, including the galactic center.

At 1 and 2 millimeters, however, the number of telescopes is much smaller than at 3 millimeters, which greatly reduces the coverage and our ability to see fine detail in the source. The small number of millimeter-capable telescopes, together with their small antenna diameters (typically 12 meters, compared with the giant older radio telescopes) and their presently limited bandwidth, which allows us to observe only the brightest heavenly objects, make VLBI observations at 1 and 2 millimeters a risky proposition.

Thankfully, this situation is rapidly changing. For example, the new Heinrich-Hertz telescope on Mount Graham near Tucson recently participated in a VLBI experiment at 1 millimeter for the first time. Even more exciting is the proposed development of the giant radio telescope known as ALMA, which conveys better than any other project the growing enthusiasm from the world's astronomical community. The Atacama Large Millimeter/Submillimeter Array (ALMA) is conceived as a radio telescope composed of 64 transportable 12-meter-diameter antennas distributed over an area 14 kilometers in extent. In the early part of 2001, representatives from Europe, Japan, and North America met in Tokyo to sign a resolution affirming their mutual intent to construct and operate this facility in cooperation with the Republic of Chile, where the telescope is to be located.

Even without the coordinated network utilizing millimeter-capable telescopes from other parts of the world, the antennas of ALMA can be pointed in unison toward a single astronomical object, and combining the signals detected by all of them with a super-fast digital signal processor, this gigantic telescope should be able to achieve an imaging detail 10 times better than that of the Hubble Space Telescope. This resolution is still not quite enough to see the black hole's shadow at the galactic center, but merged into the rest of the CMVA, ALMA will provide yet another significant addition to the fleet of independent sites spread across the globe for a transcontinental observational facility. ALMA will be built on the Andean plateau at 5,000 meters altitude near the Atacama Desert of northern Chile, and with the broad representation from the worldwide astronomical community, is considered to be the first truly global project in the history of fundamental science. The telescope is scheduled to be fully operational in 2010.

Another technical problem to overcome is that below a wavelength of 1 millimeter, Earth's atmosphere begins to absorb some of the light we are trying to sense from distant sources. The location for ALMA was selected with particular care to avoid these complications as much as possible, given that it must be built on the ground somewhere, limiting our options. The Atacama Desert provides exceptionally dry atmospheric conditions, which push the open window for observations as far into the submillimeter range as we can afford at present. The next step will be to move the platform above Earth's atmosphere entirely, and plans are underway to carry this ambitious goal to fruition in the not-too-distant future. The National Aeronautics and Space Administration (NASA) is currently sponsoring the development of a project known as the Advanced Radio Interferometry between Space and Earth (ARISE), which will be a mission consisting of one (or possibly two) 25-meter radio telescopes in highly elliptical Earth orbit. The telescope(s) would observe in conjunction with a large number of radio telescopes on the ground, using VLBI, in order to obtain the highest resolution (10-microarcsecond) images to date. By comparison, the diameter of Sagittarius A*'s shadow is predicted to be about 30 microarcseconds across (see figure 5.1), so ARISE should have no trouble seeing features less than 1/3 of its size, an astounding improvement by a factor of 100,000 over what the Hubble Space Telescope can do today.

5.4   AN X-RAY IMAGE OF THE DARK SHADOW FROM SPACE

Looking further into the future, producing images of Sagittarius A* like those we see in figures 5.1 to 5.4 may also be feasible using X-rays with a duo of powerful new NASA telescopes being proposed for flight before the year 2020, with costs estimated in the billion-dollar range. The telescopes are being developed by the University of Colorado at Boulder and NASA, and are part of the Microarcsecond X-ray Imaging Mission, or MAXIM for short. With an ultimate resolution 3 million times better than that of the Chandra X-ray Telescope, the new observatories should allow X-ray astronomers to finally peer into the dense environment within several Schwarzschild radii of the black hole. The main MAXIM mission would consist of a fleet of 33 spacecraft, each containing a relatively small telescope. But by combining the data gathered by so many separate instruments distributed over an extraordinarily large baseline in space, one may achieve a resolution of the sky about a factor of a million times better than what is currently attainable. A ground-based optical telescope with this same capability would enable us to read a newspaper on the lunar surface!

We have mentioned Sagittarius A*'s X-ray persona only briefly thus far, given that the most exciting upcoming developments are likely to occur with radio and millimeter-capable telescopes. If we are to believe that the central point in Chandra's image of the galactic center (see figure 4.1) is in fact Sagittarius A*'s X-ray alter ego, however, then most of what we've said about the black hole's shadow should apply in X-rays as well. This is true because the hot gas falling into the event horizon can play bumper cars with the photons it produces, which occasionally jump to a higher energy. Very few of the photons undergo this rude ejection from Sagittarius A*'s fold, but this should happen in sufficient numbers that Chandra could detect them. So the plasma constituting Sagittarius A*'s corona glows not only at radio and millimeter wavelengths, but also noticeably at X-ray energies as well. Mind you, the X-radiation is incredibly weak compared to that from much smaller objects, such as a pulsar, but it is detectable nonetheless, making the MAXIM mission highly desirable in the long run.

This technology has its own problems to contend with.[30] The wavelength of an X-ray is about 1,000 times smaller than that of visible light, making X-ray telescopes very difficult to build. Surface irregularities that are too small to affect visible light can easily scatter X-rays. In addition, to obtain a true focus, X-ray photons must reflect twice from very carefully figured hyperbolic and parabolic surfaces, nested concentrically in very precise formation. In this regard, the Chandra X-ray Observatory is a technological miracle, at the forefront of X-ray technology. However, there is no doubt that getting a good look at Sagittarius A* will

---

[30]This work has been spearheaded by Webster Cash and his team at the University of Colorado, in collaboration with NASA's Marshall Space Flight Center in Huntsville, Alabama. They announced their design in the 14 September 2000 issue of *Nature*.

require a leap in this capability. A single X-ray telescope in space would be impractical, since its mirror would need to be about the size of a football field—too big and too heavy to carry into Earth orbit.

Instead, MAXIM will utilize a method similar to VLBI, in which two or more telescopes are coupled in order to synthetically build an aperture equal to the separation of the individual instruments. Instead of precisely focusing X-rays with expensive mirrors onto a detector, the MAXIM team will use readily made flat mirrors to mix the photons, producing an even sharper image, similar to the way sound waves can be combined to either cancel each other out (resulting in silence) or amplify the sound when one crest adds to the other. The concept calls for the fleet of smaller telescopes to be spaced evenly in orbit around the perimeter of a circle, the diameter of which will vary from 328 feet to 3,280 feet. From there they would collect X-ray beams and funnel them to a larger telescope stationed at the hub, which could then relay the accumulated data back to Earth, several million miles away. The MAXIM constellation itself is designed to be in orbit about the Sun. Is this ambitious? Yes. Should it be done? All that needs to be said is that a separate universe awaits us at journey's end.

In the meantime, NASA's Chandra X-ray telescope has already excited the astrophysical community with a recent discovery that can only hint at what future high-precision X-ray detectors will uncover. With a simultaneous announcement in the 6 September 2001 issue of the international journal, *Nature*, and at a press conference in Washington, D.C., Fred Baganoff and his collaborators revealed the identification of a very unusual event monitored by Chandra the previous year. It was only a flash of X-rays, something seen repeatedly from many parts of the sky during the course of just one day. But this was an unusual flurry of activity indeed, since it was evidently coming from the direction of Sagittarius A*. It was over in a couple of hours, but while it lasted about 45 times as many X-rays were being emitted per second as are usually produced when Sagittarius A* is quiet. Even more startling was the observation that, during the flare, the X-ray output dropped abruptly by a factor of about 5 in less than 10 minutes, and then recovered almost as quickly. Such breathless variability is rarely seen in emissions from ponderous multimillion solar-mass objects. It is far more common, however, with stars the size of the Sun. And that's what makes it interesting.

Variability is a powerful indicator of the size of the radiating region, because nothing can travel faster than light. For an entire object's luminous output to vary considerably over a short interval, all its various parts must be able to "communicate" their changes to each other within that time. An abrupt change in the X-ray emission from Sagittarius A* over 10 minutes means that the compressed, hot, radiating gas could not have been distributed over a region bigger than the distance between Earth and the Sun, this being the distance traversed by a light beam during that interval. By comparison, the radius of Sagittarius A*'s event horizon is only about 20 times smaller. The discovery reported by Baganoff and his collaborators is therefore quite exciting, because it places the 2.6 million Suns of matter calculated from the movements of stars into a region at most 20 times larger than that predicted for this black hole by general relativity. Indications such as this give us great confidence that future high-energy missions, such as MAXIM, will therefore be able to probe the nature of space and time a mere thread above the event horizon of a giant black hole.

5.5   IMPACT ON THE GENERAL THEORY OF RELATIVITY

Throughout this book, we have documented the penetrating explanations and confirmed predictions of general relativity that have given us an appreciation of its probative power to unravel the mysteries of the universe in which we live. The bending of light was a real surprise, a major first step that over the course of a century cascaded into several notable successful tests of the theory, which has so far survived unscathed. It's not difficult to understand why it fascinates us, and why a discovery such as the binary pulsar PSR 1913+16, which confirmed the existence of gravitational radiation, could garner a Nobel prize in physics.

General relativity is almost surreal in the way it mysteriously compels us to accept truths about nature that are difficult to appreciate based solely on our everyday experiences. We experience fascination and perplexity with many other fields in science, but not quite with the profundity of this theory. Perhaps this distinction occurs because general relativity does something both enchanting and disquieting to space and time, the two main threads of our being. It folds them, twists and pulls them, and then weaves them into a single multifaceted unit.

Our curiosity is piqued by the opportunity to see what the distortions of spacetime can do. The discovery of another universe, with an alternate metric of reality, would force us to define our own; it would provide a reflection with which to measure our own world's internal constitution, to discern the weaving in the fabric of our existence. The nature of black holes has collected all these musings into one easily identifiable unknown—that's why they excite us, and haunt us even more. And Sagittarius A*, because of its size and its proximity, is our principal gateway into this uncharted territory.

The appearance of Sagittarius A*'s shadow is a firm prediction of general relativity, which mandates a unique shape and size for the region where light bending and capture are severe. We have never had a comparable opportunity to place the existence of black holes on such a firm footing. Galactic black hole binaries contain compact objects that are far too small to be imaged, and the cores of other galaxies (which we discuss in chapter 6) are too remote. At the galactic center, Sagittarius A* has just the right combination of physical attributes to place it on the verge of detectability with sub-millimeter telescopes. This coming decade may finally give us a view into one of the most important and intriguing scientific discoveries of our time. The imaging of the shadow would confirm the widely held belief that most of the dark matter concentrated in the nuclei of galaxies is a supermassive black hole, and it would provide the first direct evidence for the existence of an event horizon—the virtual membrane that keeps whole worlds apart.

On the other hand, a nondetection with sufficiently developed techniques would pose a major problem for the standard black hole paradigm. We know the mass is there, confined within a region barely the size of a planetary orbit. But if this matter is not collapsed within itself, general relativity will have broken down where gravity has a strength not sampled in any of the previous tests. Would this failure be a foreboding that quantum mechanics cannot be ignored, even at the event-horizon scale? Or perhaps it will add credence to alternate descriptions of reality, such as string theory, in which gravity is but one element in a virtual theory of everything (TOE). No one knows yet what step ought to be taken should we be faced with this unexpected, though not unprecedented, outcome.

6

QUASARS AND
GALACTIC NUCLEI

Our journey to the center of the galaxy will not be complete until
we consider its place in nature's grand scheme, for though Sagittarius A* may be unusual, it is hardly alone. It is without doubt
the most accessible object of its kind, but it was not the first to
spawn the supermassive-black-hole notion. In the course of time,
it may not even endure as the most comprehensive archetype of
this class because, although it is the easiest to see and study, its
brethren are many and their diversity is quite expansive.

Let us set our clock back to the year 1963, in the decade of
Apollo and the age of Aquarius. At the University of Texas, Roy
Kerr was developing his now-famous solution to Einstein's field
equations, describing the spacetime torqued by a spinning black
hole, while half a continent away, at Mount Palomar observatory,
Maarten Schmidt was puzzling over the nature of a starlike object
with truly anomalous characteristics. In retrospect, this conflu-
ence in time of the emergence of a theory and the uncloaking of
an object for which the theory would eventually find meaning, has
the appearance of a fated legacy.

The astronomical enigma on Schmidt's desk was the point
of light recently associated with the 273rd entry in the third
Cambridge catalog of radio sources, hence its designation as 3C
273. For centuries, such objects had gone unnoticed, appearing in the nighttime sky merely as faint stars. But with the
advent of radio astronomy in the 1940s came the gradual realization that several regions of the cosmos are very bright emit-

ters of centimeter-wavelength radiation. Just one year earlier, the British astronomer Cyril Hazard had devised an ingenious method of using lunar occultation to pinpoint the exact location of such a source. By noting the precise instant that the radio signal stopped and then re-emerged when the moon passed in front of it, astronomers could determine with which, if any, of the known visible objects in the firmament it was associated.

The observation Hazard had proposed almost didn't happen. He had arranged to make the measurements at Parkes Radio Telescope in the Australian outback, situated several hundred miles from the University of Sydney, where he was staying. On the night of the experiment, Hazard took the wrong train and missed the event entirely. However, the observatory director John Bolton and his colleagues went ahead with the plan anyway, though not without additional complications. The telescope, it turned out, couldn't tip over sufficiently to make the recording. So they cut down some intervening trees and removed the telescope's safety bolts, releasing the several-thousand-ton facility to swivel enough to catch the occultation.

The radio source captured by Parkes that night—3C 273—could be traced with precision to a single starlike object in the constellation Virgo. Although it looks like an ordinary star, this quasi-stellar radio source (hence the name *quasar*) emits a prodigious radio power, and subsequent analysis revealed that the characteristics of its optical light—basically, the colors of its rainbow—were unlike anything ever encountered before. The puzzle was solved[31] when Schmidt realized that the pattern of colors before him were simply those produced by hydrogen atoms, though with a wavelength shifted by approximately 16 percent from the value measured in the laboratory.

By then it was already known that such redshifts were indicative of large cosmic distances, so 3C 273's starlike image belied its true nature, for it had to be among the most powerful emitters of radiation in the universe. Just as the pitch of a whistle depends

---

[31] Maarten Schmidt reported his discovery in a one-page article published by *Nature* in 1963.

on how fast the train is moving toward or away from us, the shift in wavelength of light is an indicator of the speed with which its source is moving. The greater its redshift, the higher is its speed of recession. It had been known since the time of Edwin Hubble (the great astronomer for whom the Space Telescope is named) that cosmological distance scales with speed. 3C 273 appeared to be receding from us at almost 30,000 miles per second, which meant that it was 3 billion light-years from Earth. A recent image of this historic object was made with the Chandra X-ray telescope, and is shown in figure 6.1.

Even greater excitement was generated when astronomers realized that variations in the total light output of quasars such as 3C 273 were occurring over a period of only 10 to 20 months. These fluctuations implied that the size of the region producing the optical light could not exceed a few light-years since (as we saw in chapter 5) coherent fluctuations would not be established in any physical object in less time than it takes photons, which are moving at the fastest possible speed, to travel across it. For example, in order for the Milky Way to radiate as a unit, a disturbance at its nucleus would need to be transmitted to its periphery so that emitters there may "know" that they should also be radiating in unison with the center. But it would take about 50,000 years for light to carry this information, so looking at the galaxy from outside, we would not expect to see large variations in its total light output on a time scale shorter than 50,000 years.

More recent measurements with space-borne X-ray detectors have reinforced these conclusions by demonstrating that quasars are even more luminous in X-rays than they are in optical light, and that their total X-ray output can vary over a period of only hours, corresponding to a source size smaller than Neptune's orbital radius. In fact, quasars are the most powerful emitters of X-rays yet discovered. Some of them are so bright that they can be seen at a distance of 12 billion light-years. Each quasar typically releases far more energy than an entire galaxy, yet the central engine that drives this powerful activity occupies a region smaller than our solar system!

Figure 6.1 Discovered in the early 1960s, the quasar 3C 273 (the bright object in the upper-left-hand corner) was one of the first objects to be recognized as a "quasi-stellar radio source" (quasar), due to its incredible optical and radio brightness, though with very perplexing properties. Insightful analysis led astronomers to the profound conclusion that 3C 273, and the other members of its class, are not stars at all, but rather incredibly powerful objects billions of light-years away. This image was taken with NASA's Chandra X-ray telescope, using its unrivaled high resolution. Its size is about 22×22 arcseconds, which at the distance to 3C 273, corresponds to roughly $2,000 \times 2,000$ light-years. Quasars often produce high-powered jets from their cores at velocities very close to the speed of light. Here, Chandra is revealing for the first time the steady glow of X-rays from 3C 273's jet, each of whose bright knots is brighter than the entire high-energy output of many other similar objects. By now, over 15,000 similar objects have been discovered, though many tend to concentrate toward the edge of the visible universe. (Photograph courtesy of H. L. Marshall et al., NASA, and MIT)

Earlier in this book, we used a similar argument concerning the quantity of mass compressed within a region of this size to argue compellingly in favor of a supermassive black-hole nature for Sagittarius A*. Over three decades ago, physicists arrived at the same conclusion for quasars based on the apparently fantastic idea that such a small volume could be screaming across the universe with the light of 100 billion Suns. But believing that the power behind a quasar must be a supermassive black hole,[32] it was natural for people to wonder whether these objects were "naked"—deep, dark pits of matter floating aimlessly across the primeval cosmic soup—or whether they were attached to more recognizable structures in the early universe. A hint came when the strong nonstellar light from the central quasar was eliminated using mechanical or electronic means. It was possible in a few cases to detect a fuzzy haze surrounding the bright beacon, and when this light was examined carefully, it turned out to have the colors and other characteristics appropriate to a normal giant galaxy.

Finding the answers to such pressing questions concerning the nature of quasars was a principal motivation for building and deploying the Hubble Space Telescope. John Bahcall of the Institute for Advanced Study in Princeton was a strong advocate for this world-class facility. In testimony before Congress in 1978, he noted that "one needs to observe quasars with the space telescope to find out whether or not these bright point-objects, quasars, are surrounded by [the] fainter, more diffuse light of galaxies." Finally, 17 years later, Hubble did provide the answers.

6.1 THE HOST GALAXIES OF QUASARS

We now know that quasars reside in the nuclei of many different types of galaxy (see figure 6.2), from the normal to those highly disturbed by collisions or mergers. In all cases, however, the sites must provide the fuel to power these uniquely bright beacons. The lessons we have learned from Sagittarius A* at the galactic center

---

[32]See for example Salpeter (1964), Zel'dovich and Novikov (1967), and Lynden-Bell (1969).

lead us to believe that a quasar turns on when the supermassive black hole at the nucleus of one of these distant galaxies begins to accrete stars and gas from its nearby environment. The intense radiation field is emitted by the plasma on its final journey toward the event horizon. So the character and power of a quasar must depend on how much matter is available for consumption. A galaxy with a somewhat less active supermassive black hole than a quasar is called an active galaxy and its central massive core is known as an active galactic nucleus, or AGN. Our Milky Way galaxy and our neighbor, the Andromeda galaxy, are examples of normal galaxies, where the supermassive black hole has very little nearby plasma to absorb.

By now, though, we might be wondering why it is that quasars tend to shine from the edge of the visible universe, but seem to be completely absent in our vicinity. Because of their intrinsic brightness, the most distant quasars are seen at a time when the universe was but a fraction of its present age, roughly 1 billion years after the Big Bang. At the time of this writing, the distance record is held by an object with the designation SDSS 1044-0125, which was discovered from data taken with the Sloan Digital Sky Survey, coordinated by the University of Chicago and the U.S. Department of Energy's Fermi National Accelerator Laboratory. A team led by Marc Davis at the University of California at Berkeley conducted followup observations of this object, taking advantage of the unparalleled light-collecting power of the Keck Telescope in Hawaii to see its rainbow of colors and determine its redshift.

Confirming that it is indeed a quasar, these astronomers were surprised to learn that it is now the most distant object ever found in the universe, a beacon that must have been among the first objects to form, well before the birth of many galaxies. The light that we sense from it was emitted when the universe was almost 10 times smaller than it is today, very close to the limit we should be able to see anywhere. This quasar is so distant that the expansion of the universe has shifted its light, originally emitted as ultraviolet photons, through the visible portion of the spectrum and into the infrared.

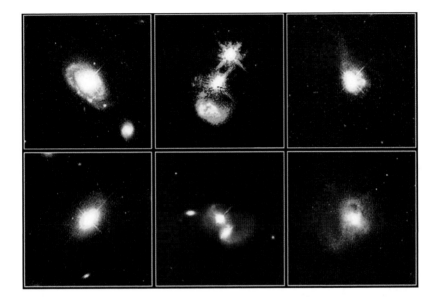

Figure 6.2 These images, taken with the Hubble Space Telescope, show three classes of hosts. In the upper image of the left column is a quasar at the core of a normal spiral galaxy 1.4 billion light-years from Earth; below it is another quasar at about the same distance, residing at the core of an elliptical galaxy. The upper photo in the central column reveals evidence of a catastrophic collision between two galaxies traveling at about 1 million miles per hour relative to each other, some 3 billion light-years from Earth. The upper object in this image is a foreground star. The quasar shines brightly in the middle, and below it sits the spiral-ring remains of a galaxy whose core was ripped out when the galaxy containing the quasar plunged through it. The current distance between the quasar and the debris of the other galaxy is about 15,000 light-years, which is roughly one-seventh the diameter of the Milky Way. In the lower photo of the central column, Hubble has captured a quasar located 1.6 billion light-years from Earth, merging with a bright galaxy to the lower right in this picture. The swirling wisps of dust and gas form a halo around these colliding objects, which are executing a sweeping dance about 31,000 light-years apart. Both of the quasars shown in the right-hand column reside within peculiar galaxies whose shape is an indication that these too have recently undergone cataclysmic collisions. Both of these are about 2 billion light-years away, and show extended arms and tails cartwheeling around the central hub. (Photographs courtesy of J. Bahcall at the Institute for Advanced Study, M. Disney at the University of Wales, and NASA)

We can understand why quasars were significantly more numerous in the early universe than they are now by noting that light travels at a finite speed, so that distant objects are seen as they existed in the past when their light was emitted. For example, we see the Sun how it was 8 minutes ago—the time it takes its radiation to reach us—and we see nearby stars as they were several years ago. When we marvel at the splendor of Andromeda's sweeping cartwheel, we are looking at light that has been traveling for roughly 2 million years. Peering to the edge of the visible universe, where the first quasars ignited, we are therefore seeing activity that occurred over 10 billion years ago. The Sloan Digital Sky Survey has been able to confirm the conclusions drawn from earlier studies, showing that the number of quasars rose dramatically from a billion years after the Big Bang to a peak around 2.5 billion years later, falling off sharply at later times toward the present.[33]

The fact that quasars turn on when it appears that fresh matter is brought into their vicinity (see figure 6.2), and then fade into a barely perceptible glimmer not long thereafter, suggests that they feed voraciously until their fuel is gone. So quasars and other types of active galactic nuclei, not to mention the relatively gentle giant at the heart of the Milky Way and in the core of Andromeda, are likely manifestations of the same phenomenon: a supermassive black hole gulping down hot gas at nearly the speed of light. Whether we see it as a quasar, or an AGN, or a Sagittarius A*-like object, probably has more to do with how much gas is present in its vicinity than anything else.

However, another remarkable recent discovery compounds the story. Back in the late 1700s, the M82 galaxy got its name when it became the 82nd entry in a systematic catalog of nebulas and star clusters compiled by the French astronomer Charles Messier (1730–1817). Recently, 220 years later, NASA's Chandra X-ray Observatory zeroed in on what appears to be a mid-sized black

---

[33]When its work is completed, the Sloan project will ultimately have surveyed one quarter of the sky and 200 million objects. About 1 million of these will be quasars, which should provide a wealth of information for statistical and evolutionary studies.

hole located about 600 light-years from its center.[34] This object packs a mass of about 500 Suns into a region no bigger than the moon. It is conceivable that this object might eventually sink to the center of M82, where it could then grow and eventually become a supermassive black hole like Sagittarius A*.

There is now an emerging class of these mid-mass black holes, 1,000 times more massive than star-sized ones like Cygnus X-1 (of Walt Disney's "The Black Hole" movie fame), yet a thousand to a million times smaller than the largest variety. They behave very much like scaled-down versions of supermassive objects found in the nuclei of the most luminous galaxies, and continue to grow as they consume matter in their vicinity. The conclusion we draw from this new category of objects is that not all the supermassive black holes in our vicinity must necessarily have begun their lives in catastrophic fashion during the quasar epoch. Some of them may have grown as malignant tumors on the substrate of existing galaxies.

6.2 THE ACTIVE NUCLEI OF OTHER GALAXIES

Directing their searchlight beams toward us from the edge of the visible universe, quasars stretch our view of the horizon as far as it can go, and awe us with their display of raw power. However, their distance of 1 billion light-years or more will always be a barrier to how much we can learn from them. At the other end of the distance scale, Sagittarius A* sits a mere 28,000 light-years away, though its frugal eating habits prevent it from displaying the full range of relativistic mass expulsion and myriad light displays more common to its distant brethren. Active galactic nuclei occupy a very useful niche between these two extremes. One of the closest and most dramatic—Centaurus A—graces the southern constellation of Centaurus, "only" 11 million light-years away (see figures 6.3 and 6.4).

---

[34]For a detailed account of this discovery, see Matsumoto et al. (2001).

Figure 6.3   The constellation of Centaurus in the southern sky contains the nearest active galaxy to Earth. At a distance of about 11 million light-years, it is roughly five times as far away as Andromeda. Its name is Centaurus A and it is the brightest radio source in this region. This image is a composite of three photographs taken by the European Southern Observatory at Kueyen, Chile, using blue, green, and red filters. Imaging this galaxy with radio telescopes, we can see two of the most spectacular jets of plasma spewing forth from the central region (see figure 6.4) in a direction perpendicular to the dark dust lanes. The nature of these jets is similar to that of the X-ray glowing stream we saw in figure 6.1, and the radio-emitting beams of particles we will view in figure 6.5. The dramatic dark band is thought to be the remnant of a smaller spiral galaxy that merged with the large elliptical galaxy. Peering toward the middle of Centaurus A, we see similar dark filaments of dust mixed with cold hydrogen gas silhouetted against the incandescent yellow-orange glow from the stars behind it. (Photograph courtesy of R. M. West and the European Southern Observatory)

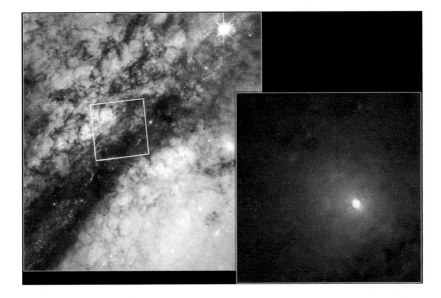

Figure 6.4 Hubble's Near Infrared Camera Multi-Object Spectrometer (NICMOS) senses infrared radiation that is significantly less sensitive to dust than optical light is (see chapter 1). Peering past the thick dust band in Centaurus A, it easily discovered a tilted disk of hot gas seen here as a (faint) white bar running diagonally across the image center. The high-speed motion of plasma in this structure, which is 130 light-years across, implies a central dark mass concentration about 300 times heavier than Sagittarius A* at the galactic center. The red blobs near the disk are glowing gas clouds that are being irradiated by the intense light from the active nucleus. The larger image to the left, whose central box is magnified at lower right, shows an enlarged view of the central region in figure 6.3, captured here with the Wide Field Planetary Camera-2 (WFPC2) on the Hubble Space Telescope. (Photographs courtesy of E. J. Schreier at the Space Telescope Science Institute, and NASA)

Like many other peculiar objects in the cosmos, Centaurus A was first observed with optical telescopes. It is now one of the most studied sources in the sky across many wavelengths, but its unique appearance had already been noted by the famous British astronomer John Herschel (1792–1871) in the mid-nineteenth century. He did not know at that time, of course, that this hauntingly beautiful structure is due to an opaque dust lane covering the central portion of the galaxy, and that the overall appearance of Centaurus A is likely the remnant of a cosmic merger between a giant elliptical galaxy, and a smaller spiral galaxy full of dust. Its origins may have mirrored the catastrophic collisions we are now witnessing with the Hubble Space Telescope in figure 6.2.

But Centaurus A is much more than this. When viewed with radio telescopes, it is one of the brightest sources in the sky, and at a distance of 11 million light-years, it is also the nearest radio galaxy. What makes it particularly relevant to our discussion of Sagittarius A*, and the other members of its class, is that the radio emission from its compact center exhibits strong activity. There are many parallels between it and our own galactic center that lead us to hypothesize the presence of a massive pointlike object in its nucleus. In addition, radio maps of Centaurus A show that its core is funneling plasma into well-collimated jets, like those seen in spectacular fashion in Cygnus A (figure 6.5), another well-known active galactic nucleus.

Astrophysicists generally agree that the intense radiation field from quasars and active galactic nuclei is produced when the infalling gas is compressed and heated to temperatures exceeding billions of degrees (see chapter 5). But unlike the situation with Sagittarius A*, which is consuming plasma at a relatively low rate, quasars and active galactic nuclei are gobbling up matter with such passion that not all of it in the gravitational whirlpool can be compressed in its entirety by the black hole. Radio and X-ray observations show that some of it is screeching away from the event horizon at nearly the speed of light, forming tightly confined streams of particles that blast through the galaxy and travel hundreds of thousands of light-years into intergalactic space. Plowing forward with very little energy loss along the way, the hot, entrained plasma continues to glow over immense distances, eventually terminating into a splash of color and woven filaments when it encounters regions of relatively high density.

For these reasons, we can sensibly expect that with appropriate measurements, we ought to be able to identify the supermassive black hole at the nucleus of Centaurus A, just as we did with Sagittarius A*. The thick dust lanes completely obscure the cen-

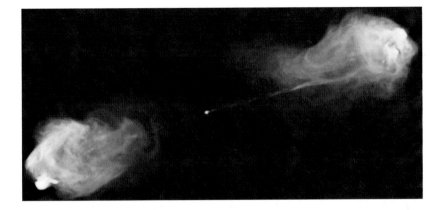

Figure 6.5 The Very Large Array (VLA) in Socorro, New Mexico, has produced many spectacular images of peculiar objects in the cosmos, but none more captivating than this glimpse of the powerful central engine and its relativistic ejection of plasma in the nucleus of the galaxy known as Cygnus A, in the constellation Cygnus—the swan. Taken at a radio wavelength of 6 centimeters, this glorious panorama reveals the highly ordered structure spanning over 500,000 light-years (more than five times the size of the entire Milky Way), fed by ultra-thin jets of energetic particles beamed from the compact radio core between them. The giant lobes are themselves formed when these jets plow into the tenuous gas that exists between galaxies. Despite its great distance from us (over 600 million light-years), it is still by far the closest powerful radio galaxy and one of the brightest radio sources in the sky. The fact that the jets must have been sustained in their tight configuration for over half a million years means that a highly stable central object—probably a rapidly spinning supermassive black hole acting like an immovable gyroscope— must be the cause of all this activity. (Photograph courtesy of C. Carilli and R. Perley, and NRAO)

tral region, but we've learned from our experience with the galactic center that radiation with longer wavelengths can seep through the intervening muck. Indeed, a combination of observations, first with the Very Large Telescope at Paranal Observatory,[35] and then

---

[35]This observation was carried out by a team of astronomers from Italy, UK, and USA, including E. Schreier (from the Space Telescope Science Institute), A. Marconi (from Arcetri Observatory), A. Capetti (from Turin Observatory), D. Axon (from the University of Hertfordshire), A. Koekemoer (from the Space Telescope Science Institute), and D. Macchetto (also from the Space Telescope Science Institute).

with an infrared detector on the Hubble Space Telescope (see figure 6.4), have finally revealed the culprit in a spectacle reminiscent of our discovery of the central point source in figure 1.4. The combination of these measurements shows that a thin gaseous disk of material is orbiting the nucleus with a blistering speed indicative of a ponderous central object. A mass of 200 million Suns is required to harness this plasma and keep it from leaving the galaxy. The astronomers who conducted these studies quickly realized that this enormous mass within the central cavity cannot be due to normal stars, since these objects would shine brightly, producing an intense optical spike toward the middle, unlike the rather tempered look of the NICMOS image shown here.

Centaurus A is too far away for us to discern yet what fraction of this condensed matter is due to dead stars as opposed to a single object, but the indications are that a supermassive black hole, with a mass possibly 300 times larger than that of Sagittarius A*, is responsible for all this activity. Aside from the compelling argument one can make based on the strong similarity between Centaurus A and the Milky Way, there is a very simple reason having to do with the existence of jets like those produced by Cygnus A in figure 6.5. At a distance of about 600 million light-years, Cygnus A is not quite as far away as many quasars, but is about 60 times more distant than Centaurus A. It is clearly much more powerful than the latter, but the two are similar in other ways, for example in producing twin jets of material moving close to the speed of light over unimaginably great distances.

When, in addition, we view the enormous cavity carved out of the universe by these energetic expulsions (see figure 6.6), we are compelled to acknowledge the sobering fact that this structure has been maintained for at least as long as it takes the streaming particles to journey from the center of the galaxy to the extremities of the giant lobes. In other words, these pencil-thin jets of relativistic plasma have retained their pristine configuration for over 1 million years! It is not surprising, therefore, that the most conservative view now maintained by physicists is that a spinning black hole is lurking at the nucleus. The axis of its spin functions like a steady rudder, an immovable gyroscope, whose direction

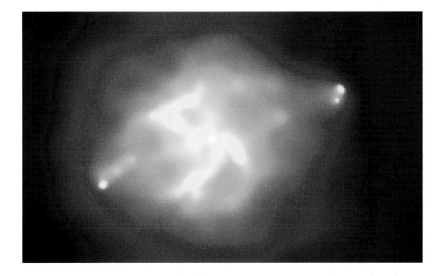

Figure 6.6 Chandra's view of the Cygnus A region is no less spectacular than that produced with the VLA (see figure 6.5), though with instruments attuned to X-rays rather than radio light, it sees different physical effects. X-rays in this source are produced primarily when relativistic particles pumped out by the central black hole cavitate the thin, hot gas pervading the medium between galaxies. This situation is very much like a balloon being inflated by gas expanding from its middle. Not surprisingly, the brightest features are those at the two termination points where the jets are impacting into the dredged up gas forming the radio lobes we saw in the previous figure. Bright bands around the equator of the enormous oval-shaped bubble are also visible, providing some evidence for the presence of material swirling toward the central black hole. (Photograph courtesy of A. S. Wilson et al. at the University of Maryland, and NASA)

predetermines the orientation of the jets. No one has produced an alternative physical description of how such a large, steady structure could otherwise be maintained. Although the definitive mechanism for how the ejection takes place is still to be worked out, most astrophysicists hold the view that the expulsion is associated with the twisting motion of magnetized plasma near the event horizon. The Kerr spacetime, which describes the dragging of inertial frames about the black hole's spin axis, provides a natu-

ral setting for establishing the preferred direction for this ejective
process.

These magnificent structures rightfully command our attention
because of the spectacular way in which they exhibit black hole
activity in their center. Other than the Milky Way, however, the
galaxy that provides us with the most compelling evidence for the
presence of a supermassive object in its core goes by the name of
NGC 4258 (i.e., the 4258th entry in the New General Catalog).
Drifting majestically some 23 million light-years from Earth, this
galaxy is easily distinguished from the others via the microwave
(maser) emission produced by water vapor in its nucleus.

In this process, light of a certain frequency is amplified when it
passes through a gas whose molecules are made to vibrate with a
higher energy than normal. In NGC 4258, the swarm of photons
produced in this fashion filter through the dust and gas enshroud-
ing the central region without any attenuation, allowing us to
probe the motion of matter deep within the core. We infer that
the water vapor is swept up by a giant whirlpool of material orbiting a strong source of gravity with velocities exceeding 1,000
kilometers per second. But these velocities occur much farther
out from the middle than those corresponding to the stars zipping
around the girth of Sagittarius A*. So the mass required to corral
this swirling disk is much greater than that lurking at the center
of our galaxy. In the core of NGC 4258, 35 million Suns of matter are condensed within a compact region no bigger than about
half a light-year across, contrasting sharply with the situation in
our neighborhood, where the same volume contains just one star.
Because of the precision with which we can measure this concentration of dark mass, we regard the nucleus of NGC 4258 as one
of the best candidates for a supermassive black hole, second only
to Sagittarius A*.

6.3   SUPERLUMINAL MOTION

In 30 to 40 known cases, the streams of particles moving at relativistic speeds away from the centers of some galaxies produce

features that move across the heavens with velocities significantly greater than that of light—a phenomenon that caused an understandable stir when it first became known.[36] The public reaction to the discovery of this apparent superluminal motion was generally one of skepticism, inducing some to refute special relativity and/or the concept of cosmological redshifts.[37] To understand what is happening, let us consider one of the most spectacular members of this class of objects—the enormous galaxy known as M87.

The Milky Way, together with Andromeda and their entourage of smaller bystanders, sit on the outskirts of the Virgo Cluster, a truly gigantic assemblage of thousands of galaxies, of which M87 is the largest. Discovered by the French astronomer Charles Messier (1730–1817) in 1781, M87's jet was first seen in 1918 by Heber Curtis[38] (1872–1942) of Lick Observatory, who described it as "a curious straight ray." M87 has become one of the most readily chosen objects for study because it is one of the nearest jet-emitting galaxies and its strong radio emission makes it an excellent target for radio telescopes. The latest high-resolution measurements (see figure 6.7) are now revealing what is happening within a mere one-tenth of a light-year from the nucleus.

Repeating the work carried out for Centaurus A—in which the Hubble Space Telescope identified in spectacular fashion not only the point source associated with the supermassive black hole in its nucleus, but also its mass—for the center of M87, astronomers

---

[36]See Zensus and Pearson (1987) for a comprehensive compilation.

[37]For an example of the discussion during this debate, see Stubbs (1971).

[38]As an interesting historical aside, Curtis was one participant in the now-famous Shapley–Curtis debate in which the featured topic was the question of whether the universe consisted of just a single giant galaxy (Shapley's position) or whether the Sun was situated near the middle of the Milky Way, a relatively small galaxy among many. A partial resolution to the debate came in the 1920s, when astronomer Edwin Hubble was successful in showing that Andromeda was much further away than the size even Shapley had attributed to the universe. Of course, the galaxy turned out to be much bigger than Curtis had allowed, and the Sun orbits well away from its center, but he was correct in his early assessment of the multigalactic structure of the universe.

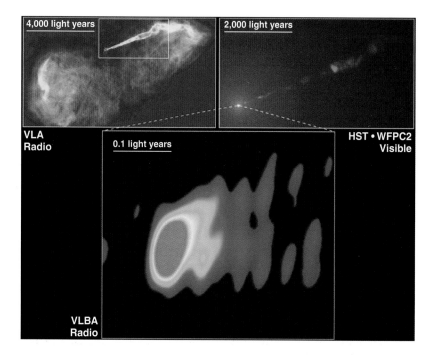

Figure 6.7 At the gravitational center of the nearby Virgo cluster of galaxies, approximately 50 million light-years away, sits a rather ordinary-looking elliptical, known as M87. Even back in the early 1900s, astronomers could see a "ray of light" protruding from its core. Streaming out from the center of M87 like a cosmic searchlight is one of nature's most amazing phenomena—a black-hole-powered jet of sub-atomic particles traveling at nearly the speed of light. The source of this grand display is a powerful central object with a mass 1,000 bigger than that of Sagittarius A*. The light that we see is produced by electrons twisting along magnetic field lines in the jet, a process that gives it the bluish tint in the upper-right-hand image. This sequence of photographs shows progressively magnified views. *Top left:* a VLA image showing the full extent of the jets and the blobs at the termination points; *top right:* a visible light image of the giant elliptical galaxy M87, taken with NASA's Hubble Space Telescope; *bottom:* a Very Large Baseline Array (VLBA) radio image of the region surrounding the black hole. The VLBA is a network of 10 radio telescopes utilizing very large baseline interferometry to discern fine detail in the source. This breathtaking view shows how the extragalactic jet is formed into a narrow beam within a few tenths of a light-year from the nucleus (the red region is only a tenth of a light-year across). This distance corresponds to only 100 Schwarzschild radii for a black hole of this mass! (Photographs courtesy of the National Radio Astronomy Observatory and the National Science Foundation [top left and bottom], and J. Biretta at the Space Telescope Science Institute, and NASA [top right])

have now shown that the object responsible for the display in the panels of figure 6.7 has a mass of 3 billion Suns (about 1,000 times bigger than Sagittarius A*!), concentrated into a volume no larger than our solar system. We therefore know (from the definition of a Schwarzschild radius—see chapter 3) that the red region evident in the bottom panel of this figure is smaller than about 100 Schwarzschild radii. In other words, with an object like the supermassive black hole at the center of M87, we are literally on the verge of seeing how these splendid jets form not too far from the event horizon.

But perhaps the best reason to marvel at this unusual galaxy is that radio telescopes (principally the VLBA), as well as the Hubble Space Telescope, have independently measured the motion of concentrations of material near the base of M87's jets, and have shown this plasma to be moving at apparent speeds six times greater than that of light (see figure 6.8). It was in fact this type of situation with quasars back in the 1960s and 1970s that led to a need for much higher resolution in the observations to see what was happening near the supermassive object. The desire to overcome the observational shortcomings motivated groups in Canada, the United States, England, and the former Soviet Union to develop Very Long Baseline Interferometry (VLBI), in which a global network of independent radio telescopes can merge their detected signals to provide exquisitely finer detail in the source than any of them could produce individually.

The phenomenon of superluminal motion is intriguing, but it doesn't have to do with the actual propagation of matter at these velocities. It's an artifact of the finite *and* constant speed of light. It was Martin Rees (1966) who first proposed that these effects might be associated with relativistically expanding shells of matter. In later refinements of this basic picture, it became clear that the underlying sources for the observed radiation were quasar jets consisting of beams of relativistically moving particles.[39]

The superluminal effect, it turned out, could be understood comfortably *within* the confines of special relativity. It has to do

---

[39]See Blandford, McKee, & Rees (1977).

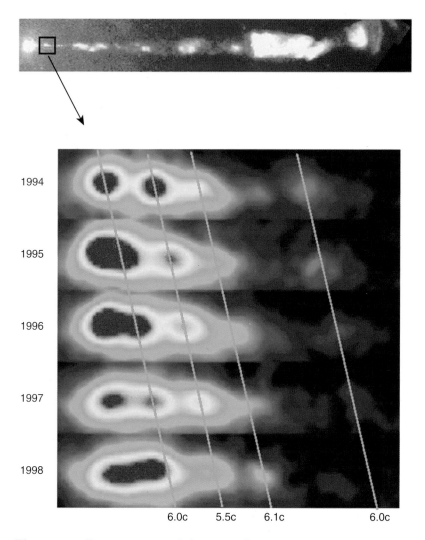

Figure 6.8 Zooming in toward the origin of the optical jet seen in the upper-right-hand panel of figure 6.7, Hubble was able to capture the motion of some of the knots expelled by the central black hole over a period of about four years. The apparent velocity of this plasma across the plane of the sky is roughly six times the speed of light. The upper panel in this figure shows the location (box) within the 5,000-light-year-long jet where this "superluminal" motion was detected. The bottom panels show the sequence of images produced by Hubble over the four-year period. The slanted lines show the moving features, labeled with their corresponding speed in units of the velocity of light, c. All these images were made with the Faint Object Camera on the Hubble Space Telescope. (Image courtesy of J. Biretta at the Space Telescope Science Institute, and NASA)

with the fact that when a source of light is moving toward us, it catches up with the radiation it is emitting, so that changes we see in its complexion appear to be happening over a smaller interval of time. Since it then appears that the distance was covered over a shortened duration, we infer a greater velocity than the object actually possessed.

Think of it this way. Suppose you and a friend are playing catch across a tall hedge, so that you could see the ball thrown over it, but not the other person. And suppose that your friend now says that she will throw two balls, in quick succession, and that based on the interval of time between them arriving in your hands, you should calculate how fast she was moving when she threw them. You know that she is going to run 10 meters between the two throws, but what you don't realize is that she has decided to run straight toward you. So she starts her run, throws the first ball, and waits until she has covered 10 meters before throwing the second.

The point is that she has moved in the same direction as the first ball (and she has partially caught up with it) so that after the second ball is thrown, the time between its arrival in your hands and that of the first is less than the time she actually waited between the two throws. The velocity you infer for her is therefore greater than what she can muster because you think she has covered the 10 meters in less time than it actually took her to do so.

Notice that this works because she is moving toward you. If she were instead running parallel to the hedge, your measured interval of time would be the same as hers and you would both agree on her velocity. By extrapolation, we infer that the observation of superluminal motion in quasars and active galactic nuclei must imply that their jets are aligned more or less toward us. What we see in images like those in figure 6.7 is the jet *projected* onto the plane of the sky. Alien eyes viewing these beams of plasma from a different angle would see them extending much further out than what is apparent to us here.

6.4   THE SUPERMASSIVE BLACK HOLE MENAGERIE

The objects we have looked at in this chapter are each special for one reason or another, either because it was the first quasar ever discovered (3C 273), or because it is the nearest radio galaxy (Centaurus A), or because it is the active galactic nucleus with arguably the most spectacular pair of jets and radio lobes (Cygnus A), or perhaps because it offers us the best opportunity of learning how the relativistically moving plasma is ejected by the central black hole (M87). But there are billions of galaxies throughout the cosmos extending, as we have seen, to the edge of the visible universe, where the evidence for the existence of supermassive black holes first appears. How many brethren, we wonder, does Sagittarius A* actually have?

Many astronomers suspect that almost every large, normal galaxy harbors a supermassive black hole at its center, a hypothesis for which supporting evidence is growing gradually. In a recently completed survey of 100 nearby galaxies using the Very Large Array radio telescope in New Mexico,[40] followed by closer scrutiny with the Very Long Baseline Interferometry array, at least 30 percent of this sample showed tiny, compact central radio sources bearing the unique signature of the quasar phenomenon. Their faint glimmer appears to be the relic signature of headier days long past, when some of these objects may have been the powerful beacons shining from the universe's epoch of structure formation.

In another notable development, NASA's Chandra X-ray telescope appears to have settled a long-standing puzzle, dating back to the early 1960s. The universe, it turns out, is aglow with the faint murmur of diffuse X-ray light, which fills the entire sky. Using a deep exposure of a selected piece of the heavens,[41] Chandra

---

[40]This work was carried out by A. Wilson from the University of Maryland and his collaborators.

[41]This work was reported by R. Mushotzky, from the NASA Goddard Space Flight Center, and his collaborators, in the international journal *Nature*, in 2000.

was able to resolve at least 80 percent of the X-ray glow into myriad individual point sources, suggesting an extrapolated total number of about 70 million across the entire sky. In follow-up studies of some of these objects, using telescopes to sense their radiation at other wavelengths, the researchers who carried out these observations concluded that some are relatively normal galaxies with dust-veiled, X-ray emitting nuclei—the signature of a central black hole.

Others are very distant quasars, too faint to shine brightly like those in figure 6.2. And the rest are either unknown or not related to supermassive black holes. The picture that emerges from surveys such as these is that at least some of the supermassive black holes must have formed very early, perhaps even before the primordial age when galaxies were first coagulating from the fragmentary gas clouds condensing out of the Big Bang. Some of them may be true fossils from the cosmological "Dark Age," the period extending over several hundred million years between the cooling of the Big Bang and the epoch of star formation.

Still, the more conservative among us would argue that these distant points of light are highly suggestive of the presence of supermassive black holes, but that this conclusion is only tentative until we actually see them, or deduce their presence from more compelling evidence, as we did in chapter 2 for Sagittarius A*. Spearheading this approach, John Kormendy at the University of Texas and Doug Richstone at the University of Michigan, and their collaborators, have set about the task of meticulously assembling the clues required to complete the detective work for as many of these objects as is currently feasible. At the time of this writing, direct measurements of supermassive black holes have been made in at least 38 galaxies, based on the large rotation and random velocities of stars and gas near their centers. These objects are all relatively nearby because these direct methods don't work unless we can see the individual stars in motion about the central source of gravity.

Curiously, though, none have been found in galaxies that lack a central bulge. Galaxies come in two basic types, those that are flat, spinning disks, and others that have more nearly spherical bulges that rotate only a little and are composed mostly of randomly moving stars. Many galaxies, such as the Milky Way and Andromeda, consist of a disk *and* a bulge in the middle. Galaxies that contain only a bulge but no disk are called ellipticals. So far, astronomers have found a supermassive black hole in every galaxy observed that contains a bulge component, but none in those that only possess a disk. In the future, we will have a better understanding of why this happens. Perhaps a collision, like those seen in figure 6.2, is required to make such an object in every case, during which the highly ordered cartwheel structure of the bulgeless galaxies is disrupted.

For the more distant active galactic nuclei, and certainly for quasars, we have no hope of pinning down the central culprit using techniques such as these. For them, our sleuthing must take on added sophistication, employing even less direct methods, which unfortunately are also subject to potential error and misinterpretation. One technique that is gaining in popularity is known as "reverberation mapping," whose basis is quite simple to understand. Since the central object in these sources is so variable, the nearby gas is irradiated with light of rapidly changing intensity. The numerous clouds of plasma that orbit each supermassive black hole in turn re-radiate the energy they absorb, but with colors indicative of their elemental content and other physical attributes. As the black hole varies its rate of energy output, the brightness of the radiation from the orbiting clouds varies as well. But because light travels at a finite speed, the variations seen from the clouds are detected some time after those associated with the central engine. This time difference tells us how far the clouds are orbiting from the center, and together with their measured speed, this information then points to the required central mass to keep them in harness.

Though not as accurate as methods employed with galaxies much closer to us, these measurements nonetheless serve the highly useful purpose of confirming our suspicion that most of the quasar activity is taking place within a very small volume, no bigger than a few light-years across. So even if we can't be certain of their exact mass, we know quite assuredly that an object much like Sagittarius A* beats at the heart of these distant beacons as well.

6.5 FUTURE HORIZONS

In this book, we have focused exclusively on the mystery of *supermassive* black holes, and Sagittarius A* in particular, because they are the most likely to succumb to our incessant attempts to actually see them in the foreseeable future. But observational science advances on many fronts, and even the comparatively tiny versions of these objects are ripe for bold new initiatives to tackle some of the most intriguing questions in physics today. What actually does happen to matter as it scuttles across the event horizon? The Hubble Space Telescope, gazing at one of the first black holes ever discovered, may have started this important assault by finding the first traces of evidence that matter is disappearing through the membrane of no return. The name of this 10-solar-mass object is Cygnus X-1, and it is feasting on its binary stellar companion some 6,000 light-years from Earth in the constellation of the swan.

There is no hope for Hubble to actually see Cygnus X-1's event horizon, but instead it measured the chaotic fluctuations in the intensity of ultraviolet light emitted by the seething plasma orbiting the black hole just above the precipice. Occasionally, blobs of gas break free from this whirlpool and spiral into the dark pit. Hubble managed to see the so-called "dying pulse train" associated with two such events, the rapidly decaying, precisely sequential flashes of radiation emitted by the infalling gas.[42] This sequence of pulses is easily recognized as the signature of matter falling so close to the event horizon that its light rapidly dimmed as it was stretched by the strong gravitational field to ever-longer wavelengths. If the end point of its plunge were a hard surface, as expected for a neutron-star accretor, the blob of gas would in-

---

[42]The investigator who mined the Hubble archives for this interesting result is J. F. Dolan from NASA's Goddard Space Flight Center in Greenbelt, Maryland.

stead have brightened upon crashing. But Hubble saw the gas crossing over into a twilight-zone realm where time slows down, and eventually switches roles with space.

The clinching piece of evidence for this interpretation is that the pulsation of the blob—an effect caused by the black hole's intense gravity—also shortened as it fell closer to the event horizon, a result of the redshift due to the endless stretching of time. Each of these events spanned an interval barely 0.2 second long. Imagine what we could learn from this object and others like it by splicing up the time frames 10-fold, or a 100-fold, and improving the coverage so that we could see many of these events. We can hardly wait for the next generation of telescopes whose temporal and spatial resolving capabilities will dwarf those of the Hubble and Chandra, and the rest of the current fleet of instruments at our disposal.

As serious black-hole hunters, though, our thoughts inevitably return to the prey of choice, Sagittarius A*, for it alone is primed for a direct imaging of its maw within this decade. Lying on the shores of Port Douglas some day, you will re-awaken from this galactic reverie to hear in the distance a rustling of palm trees lining the beach on the edge of Earth's oldest living rainforest. The grandeur of the Milky Way above you will re-infuse your senses with the spirit of wonder and anticipation. In Jules Verne's description, Professor Hardwigg's journey to the center of our planet "created an enormous sensation throughout the civilized world." One can only imagine what hyperbole he may have used in celebration of this excursion to the center of the galaxy.

REFERENCES

Alexander, G. H., *The Leibniz-Clarke Correspondence*, Manchester University Press, 1956.

Anderson, J. L., *Principles of Relativity Physics*, Academic Press, 1967.

Backer, D. C., and R. A. Sramek, "Proper Motion of the Compact, Nonthermal Radio Source in the Galactic Center, Sagittarius A*," *Astrophysical Journal* 524 (1999), pp. 805–815.

Balick, B., and R. L. Brown, "Intense sub-arcsecond structure in the galactic center," *Astrophysical Journal* 194 (1974), pp. 265–270.

Blandford, R. D., C. F. McKee, and M. J. Rees, "Superluminal Expansion in Extragalactic Radio Sources," *Nature* 267 (1977), pp. 211–216.

Bromley, B., F. Melia, and S. Liu, "Polarimetric Imaging of the Massive Black Hole at the Galactic Center," *Astrophysical Journal Letters* 555 (2001), pp. L83–L87.

Cash, W., A. Shipley, S. Osterman, and M. Joy, "Laboratory Detection of X-ray Fringes with a Grazing-incidence Interferometer," *Nature* 407 (2000), pp. 160–162.

Chadwick, H., *The Confessions of St. Augustine,* Oxford University Press, 1998.

Cunningham, C. T., and J. M. Bardeen, "The Optical Appearance of a Star Orbiting an Extreme Kerr Black Hole," *Astrophysical Journal* 183 (1973), pp. 237–264.

Damour, T., in *Three Hundred Years of Gravitation*, S. W. Hawking and W. Israel, editors, Cambridge University Press, 1987, pp. 128-198.

Drake, S., *Cause, Experiment, and Science*, Chicago University Press, 1981.

Duschl, W. J., and H. Lesch, "The Spectrum of Sagittarius A* and Its Variability," *Astronomy and Astrophysics* 286 (1994), pp. 431–436.

Falcke, H., P. L. Biermann, J. D. Wolfgang, and P. G. Mezger, "A Rotating Black Hole in the Galactic Center," *Astronomy and Astrophysics* 270 (1993), pp. 102–106.

Falcke, H., F. Melia, and E. Agol, "Viewing the Shadow of the Black Hole at the Galactic Center," *Astrophysical Journal Letters* 528 (2000), pp. L13–L17.

Gillispie, C. C., *Pierre-Simon Laplace, 1749-1827: A Life in Exact Science*, Princeton University Press, 1997.

Gwinn, C. R., R. M. Danen, et al., "The Galactic Center Radio Source Shines Below the Compton Limit," *Astrophysics Journal Letters* 381 (1991), pp. L43–L46.

Guthrie, W.K.C., *A History of Greek Philosophy, Volume I: The Earlier Pre-Socratics and the Pythagoreans,* Cambridge University Press, 1962.

Hawking, S. W., "Black Hole Explosions?" *Nature* 248 (1974), p. 30.

Hollywood, J. M., and F. Melia, "The Effects of Redshifts and Focusing on the Spectrum of an Accretion Disk in the Galactic Center Black Hole Candidate Sgr A*," *Astrophysics Journal Letters* 443 (1995), pp. L17–L21.

———, "General Relativistic Effects on the Infrared Spectrum of Thin Accretion Disks in AGNs; Application to Sgr A*," *Astrophysics Journal Supplements* 112 (1997), pp. 423–455.

Hollywood, J. M., F. Melia, L. M. Close, D. W. McCarthy Jr., and T. A. DeKeyser, "General Relativistic Flux Modulations in the Galactic Center Black Hole Candidate Sgr A*," *Astrophysics Journal Letters* 448 (1995), pp. L21–L25.

Kerr, R. P., "Gravitational Field of a Spinning Mass as an Example of Algebraically Special Metrics," *Physical Review Letters* 11 (1963), pp. 237–238.

Khokhlov, A., and F. Melia, "Powerful Ejection of Matter from Tidally Disrupted Stars Near Massive Black Holes and a Possible Application to Sgr A East," *Astrophysics Journal Letters* 457 (1996), pp. L61–L64.

Kormendy, J., and D. Richstone, "Inward Bound—The Search for Supermassive Black Holes in Galactic Nuclei," *Annual Reviews of Astronomy and Astrophysics* 33 (1995), pp. 581–624.

Lubowich, D. A., J. M. Pasachoff, et al., "Deuterium in the Galactic Centre as a Result of Recent Infall of Low-metallicity Gas," *Nature* 405 (2000), pp. 1025-1027.

Lynden-Bell, D., "Galactic Nuclei as Collapsed Old Quasars," *Nature* 223 (1969), p. 690.

Lynden-Bell, D., and M. J. Rees, "On Quasars, Dust and the Galactic Centre," *Monthly Notices Roy. Astron. Soc.* 152 (1971), p. 461.

Mach, E., *The Science of Mechanics,* trans. by T. J. McCormack, 2nd ed., Open Court Publishing, 1893.

Matsumoto, H., et al., "Discovery of a Luminous, Variable, Off-Center Source in the Nucleus of M82 with the Chandra High-Resolution Camera," *Astrophysics Journal Letters* 547 (2001), pp. L25–L28.

Maxwell, J. C., *Treatise on Electricity and Magnetism,* Vol II, Dover Publications, 1954.

Melia, F., "An Accreting Black Hole Model for Sagittarius A*," *Astrophysics Journal Letters* 387 (1992), pp. L25–L29.

Melia, F., and H. Falcke, "The Supermassive Black Hole at the Galactic Center," *Annual Reviews of Astronomy and Astrophysics* 39 (2001), pp. 309–352.

Michelson, A. A., and E. W. Morley, "On the Relative Motion of the Earth and the Luminiferous Ether," *American Journal of Science* 34 (1887), pp. 333–345.

Misner, C. W., K. S. Thorne, and J. A. Wheeler, *Gravitation*, W.H. Freeman, 1973.

Mushotzky, R. F., L. L. Cowie, A. J. Barger, and K. A. Arnaud, "Resolving the Extragalactic X-ray Background," *Nature* 404 (2000), pp. 459–464.

Narayan, R., I. Yi, and R. Mahadevan, "Explaining the Spectrum of Sagittarius A* with a Model of an Accreting Black Hole," *Nature* 374 (1995), pp. 623–624.

Newton, I., *Philosophiae Naturalis Principia Mathematica,* trans. by Andrew Motte, revised and annotated by F. Cajori, University of California Press, 1966.

Penrose, R., "Gravitational Collapse, the Role of General Relativity," *Rivista Nuovo Cimento* 1 Numero Speciale (1969), p. 252.

Pound, R. V., and G. A. Rebka, "The Apparent Weight of Photons," *Physical Review Letters* 4 (1960), pp. 337–341.

Rees, M. J., "Appearance of Relativistically Expanding Radio Sources," *Nature* 211 (1966), pp. 468–470.

——, "Tidal Disruption of Stars by Black Holes of 10^6 to 10^8 Solar Masses in Nearby Galaxies," *Nature* 333 (1988), pp. 523–528.

Reid, M. J., A.C.S. Readhead, R. C. Vermeulen, and R. N. Treuhaft, "The Proper Motion of Sagittarius A*. I. First VLBA Results," *Astrophysics Journal* 524 (1999), pp. 816–823.

Salpeter, E. E., "Accretion of Interstellar Matter by Massive Objects," *Astrophysics Journal* 140 (1964), pp. 796–800.

Schmidt, M., "3C 273: A Star-like Object with Large Red-shift," *Nature* 197 (1963), p. 1040.

Serabyn, E., J. H. Lacy, C. H. Townes, and R. Bharat, "High-resolution Forbidden NE II Observations of the Ionized Filaments in the Galactic Center," *Astrophysics Journal* 326 (1988), pp. 171–185.

Shai, A., M. Livio, and T. Piran, "Tidal Disruption of a Solar-Type Star by a Supermassive Black Hole," *Astrophysics Journal* 545 (2000), pp. 772–780.

Strohmayer, T. E., "Discovery of a 450 Hz QPO from the Microquasar GRO J1655-40 with RXTE," *Astrophysics Journal Letters* 552 (2001), pp. 49–53.

Stubbs, P., "Red Shift Without Reason," *New Scientist* 50 (1971), pp. 254–255.

Thorne, K. S., *Black Holes and Time Warps: Einstein's Outrageous Legacy,* Norton, 1995.

Thorne, K. S., R. H. Price, and D. A. Macdonald, *Black Holes: The Membrane Paradigm,* Yale University Press, 1986.

Verne, J., *Journey to the Center of the Earth,* Penguin, 1965.

Wald, R., *General Relativity,* University of Chicago Press, 1984.

Weber, J., "Evidence for Discovery of Gravitational Radiation," *Physical Review Letters* 22 (1969), pp. 1320–1324.

Weinberg, S., *Gravitation and Cosmology: Principles and Applications of the General Theory of Relativity,* Wiley, 1972.

Westfall, R., *Never at Rest,* Cambridge University Press, 1981.

Wheeler, J. A., *Journey into Gravity and Spacetime,* Freeman, 1999.

Wright, M.C.H., A. L. Coil, et al., "Molecular Tracers of the Central 12 Parsecs of the Galactic Center," *Astrophysics Journal* 551 (2001), pp. 254–268.

Zel'dovich, Ya. B., and I. D. Novikov, "The Hypothesis of Cores Retarded During Expansion and the Hot Cosmological Model," *Soviet Astronomy* 10 (1967), p. 602.

Zensus, J. A., and T. J. Pearson, *Superluminal Radio Sources,* Cambridge University Press, 1987.

INDEX